Statistics and Computing

Series Editors
J. Chambers
D.J. Hand
W. Härdle

For other titles published in this series, go to
http://www.springer.com/series/3022

A SAS/IML Companion
for Linear Models

Jamis J. Perrett

Texas A&M University

Springer

Jamis J. Perrett
Department of Statistics
Texas A&M University
College Station
TX 77845-3143
USA
jamis@stat.tamu.edu

Series Editors
J. Chambers
Department of Statistics
Sequoia Hall
390 Serra Mall
Stanford University
Stanford, CA 94305-4065

D.J. Hand
Department of Mathematics
Imperial College London
South Kensington Campus
London SW7 2AZ
United Kingdom

W. Härdle
Institut für Statistik und Ökonometrie
Humboldt-Universität zu Berlin
Spandauer Str. 1
D-10178 Berlin
Germany

ISSN 1431-8784
ISBN 978-1-4419-5556-2   e-ISBN 978-1-4419-5557-9
DOI 10.1007/978-1-4419-5557-9
Springer New York Dordrecht Heidelberg London

Library of Congress Control Number: 2009941075

Springer is part of Springer Science+Business Media (www.springer.com)

Preface

This companion is designed for researchers who wish to perform linear models analysis using the definitional formulas. It can serve as a supplemental text for students in a college-level theoretical or applied linear models course using The SAS System (SAS) for computations. This book can also serve as the principal text for a special topics course in a graduate program in statistics with linear models theory and applications courses as prerequisites. It may also serve as a useful reference for statistical analysts who wish to use the SAS/IML product to work out the formulas for new experimental methods of data analysis.

A linear models course can be taught from different approaches. Some of these approaches are more theoretical and focus on computation and derivation of the linear algebra formulations. Such a course might consist of theorem, proof, theorem, proof, theorem, proof, theorem, lemma, proof, etc. Some theoretical approaches focus on geometric interpretations of the linear model using projections. Yet other approaches are more applied and focus on the computer-aided applied analysis using high-level computational procedures such as PROC GLM, MIXED, REG, etc. in SAS, with little time spent with the analytic formulas.

This companion illustrates the theoretical linear algebra approach to teaching linear model analysis. This companion does not teach linear models concepts, but demonstrates how SAS/IML can be used to evaluate numerical linear model problems. It is assumed that the reader does not have any experience using SAS/IML, but is familiar with DATA steps and basic PROC steps in the SAS system.

All the SAS examples given in this companion are self-contained, and may be executed as written, without additional programming. In most cases, there are other SAS procedures that are more appropriate to use in the analysis of linear models than the IML procedure. In many cases, the analysis will be performed in both IML and another, more appropriate SAS procedure. This is done for the following reason: The IML approach is directed at leaning and applying the linear models formulas. The more "canned" procedures like REG and GLM do not allow students to see the connection between formulas and the procedures. However, because REG and GLM are more efficient and numerically accurate than SAS/IML in many applications of the linear model, they will also be briefly demonstrated. The canned procedures (such as REG and GLM) are considered the standard for analysis and are provided to demonstrate that the results obtained using the IML procedures are the same as those using the canned procedures. This allows students to go from written analytic formula in say mixed models analysis to IML implementation of that same analytic formula to high level analysis using PROC MIXED. Adding the step of PROC IML implementation provides the missing link in the learning process.

The topics selected for this companion are the topics the author found to be useful in the academic learning of the analysis of linear models. There are many additional topics useful in the study of linear models that were not included. However, this should give the reader a complete treatise for a course on the subject.

Exercises have been included to aid the learner. Some exercises were developed by withholding IML implementation steps for the reader to work through. Other exercises are developed for perhaps repetitive application of concepts introduced in the discussion.

The examples presented take advantage of the most recent advances in SAS/IML and are current as of SAS version 9.2. However, most of the examples will work in earlier versions of SAS.

Contents

List of Examples

1.4 Formulas – From Paper to Computer

IML is good at automating text book formulas. The code written in IML closely
matches hand-written formulas. For example, the line of code used to compute the
equation

$$H = X(X'X)^{-1}X'$$

would be of the form

```
h=x*INV(x'*x)*x';.
```

Notice that where named functions are used they are often indicative of the mathematical symbol they represent. The above equation uses a negative one superscript to
represent the inverse of a matrix (see section 5.1.4). Because superscripting is not a
part of ANSI code, it cannot be replicated in SAS code. Thankfully, the named function that takes the place of this inverse is called INV. The benefit of the code written
in IML closely matching hand-written formulas is that researchers and students can
convert formulas (that can be written in terms of matrices) they are working with on
paper to IML code with relative ease.

1.5 Getting Started

Because IML is an interactive procedure, using IML starts by submitting the code,

```
PROC IML;
```

to invoke the IML computing environment. Once in the IML computing environment, the user can submit one line of code at a time, or several at a time, and SAS
processes the requests as they are submitted without leaving the IML computing
environment. The results of the submitted code are sent to the LOG or the OUTPUT
appropriately, and the user may continue working with IML until desiring to leave
that environment, at which point the statement

```
QUIT;
```

is submitted and SAS exits the IML computing environment and returns to the SAS
computing environment. A given program may consist of multiple PROC IML steps,
along with other SAS PROC steps and DATA steps.

   IML code may be submitted in sections of one or more lines. This allows the
user to inspect the results of each submitted line of code without having to leave the
IML computing environment and later return. The following example illustrates the
interactive nature of IML. The code is submitted one line at a time for the purpose
of illustration.

Example 1.1 – Computing the sum of the elements of matrix A (interactively):

This example computes the sum of the elements in the matrix **A**. The purpose is
to show the interactive nature of IML. Right angle brackets (>) precede submitted

statements. Responses from SAS, that appear in the LOG, follow. The framed text
appears in the OUTPUT.

```
>    PROC IML;
IML Ready
>    a={1 2 3,2 3 1,2 4 6};
a={1 2 3,2 3 1,2 4 6};
>    sum_a=SUM(a);
sum_a=SUM(a);
>    PRINT A SUM_A;
>    QUIT;
QUIT;
```

a			sum_a
1	2	3	24
2	3	1	
2	4	6	

```
NOTE: Exiting IML.
```

The statement

```
PROC IML;
```

starts the PROC IML step and invokes the IML computing environment. The
statement

```
a={ 1 2 3,2 3 1,2 4 6};
```

creates the 3×3 matrix **A**. The statement

```
sum_ a=SUM(a);
```

computes the sum of the elements of the matrix **A**. The statement

```
PRINT A SUM_A;
```

prints the matrix A and the sum, SUM_A to the output. The statement

```
QUIT;
```

finishes the PROC IML step and exits the IML computing environment.

1.6 What is an IML Matrix?

IML is a matrix language. For the purpose of IML, a matrix is a collection of numbers or alphanumeric characters that has its own space in a group of one or more
rows and columns. The way these numbers or characters are placed in the matrix
defines characteristics about the matrix. A *vector* is a matrix with only one row or

one column. A vector can be identified as either a *row vector* (a matrix with multiple columns but only one row) or as a *column vector* (a matrix with multiple rows but only one column). A *scalar* is a matrix with only one element. The following two examples give samples of numeric and character matrices.

1.6.1 A Numeric Matrix

The numbers 92, 95, 83, 87, and 96, which represent the scores of five students on an exam, can be placed in five rows of a single column in the following manner:

$$\mathbf{SCORES} = \begin{bmatrix} 92 \\ 95 \\ 83 \\ 87 \\ 96 \end{bmatrix}$$

SCORES can be referred to as a matrix, but because it contains only one column, it is more appropriately referred to as a column vector.

1.6.2 A Character Matrix

The following matrix, **CLASS_A**, is a matrix containing five rows and two columns. The first column represents the names of the students, and the second column represents the students' scores on an exam.

$$\mathbf{CLASS\_A} = \begin{bmatrix} \text{Rob} & 92 \\ \text{joe} & 95 \\ \text{Erich} & 83 \\ \text{Lisa} & 87 \\ \text{Mike} & 96 \end{bmatrix}$$

There are innumerable ways of forming matrices with characters and numbers, and just as many applications for such. Though there is one numeric column in the matrix **CLASS_A**, the fact that it has non-numeric characters in the other column makes this a character matrix. In fact, a single non-numeric character in a matrix makes the entire matrix non-numeric (or "character" as it is being labeled here). As mathematical operations cannot be conducted on character matrices, character matrices are primarily used to contain labels for the elements of numeric matrices or as a method of organization of data as an alternative to the method of organization of data found in SAS data sets.

1.6.3 A Linear System of Equations

As mentioned before, linear algebra is conducted using numeric matrices. Numeric matrices are often used in mathematics as a way to simplify the solutions to multiple or complex equations.

Example 1.2 – A linear system of equations:

Consider the following: three algebraic equations with three unknown variables.

$$5a + 3b - 2c = 5$$
$$3a + 4b - 3c = 2 \qquad\qquad (1.1)$$
$$4a + 2b - 5c = -7$$

Using a series of steps, these three equalities can be used to solve for the three unknown values. Alternately, the numbers on the left side of the equality can be organized into the rows and columns of a numeric matrix and the numbers of the right side of the equality can be organized into a numeric column vector. Using linear algebra, the resulting matrix and column vector can be combined in a single equation and solved for the values of the unknown variables. Not only that, but this methodology can be used for any number of unknown variables that have the same number of associated linearly independent equations. The matrix representation of the set of three equations is

$$AX = C,$$

where

$$A = \begin{bmatrix} 5 & 3 & -2 \\ 3 & 4 & -3 \\ 4 & 2 & -5 \end{bmatrix}$$

$$X = \begin{bmatrix} a \\ b \\ c \end{bmatrix}$$

$$C = \begin{bmatrix} 5 \\ 2 \\ -7 \end{bmatrix}$$

The idea is to solve for the values of **X**. The solution can be found using the following method:

$$X = A^{-1}C \qquad\qquad (1.2)$$

The solution to the set of three equations (1.1) is a=1, b=2, and c=3. This solution can be computed using the following SAS code:

```
PROC IML;
   a={5 3-2,
      3 4 -3,
      4 2 -5};
   c={5,2,-7};
   x=INV(a)*c;
   PRINT x;
QUIT;
```

The statements

```
   a={5 3 -2,
      3 4 -3,
      4 2 -5};
```

and

```
   c={5,2,-7};
```

define the matrix **A** and column vector **C**, respectively. The statement

```
   x=INV(a)*c;
```

defines the equality found in Equation1.2. The statement

```
   PRINT x;
```

prints the column vector **X** to the output. The result is the following.

x

1

2

3

1.7 Notation

It is customary to refer to a matrix containing five rows and two columns (as with the matrix **CLASS_A** in Example 1.3) as a "5 × 2 matrix." The number of rows is always mentioned before the number of columns. For this text, inline text matrices and vectors are labeled with uppercase text and bold face. Within the code, matrices and vectors are labeled with a lowercase text and regular face. SAS is not case sensitive. So, in practice matrix, vector, and variable names can be all uppercase, all lowercase, or mixed case.

1.8 Viewing the Contents of a Matrix

The IML procedure has a feature that allows you to print the contents of a matrix to the output. This can be useful to verify a matrix was created correctly and to view the matrix resulting from an equation.

Printing a matrix to the output window is quite simple. It only requires the command

```
PRINT [matrix_name];
```

from within the IML procedure. Here is an example of how this is done.

Example 1.3 – Viewing a matrix:

This example creates a scalar, **X**, containing the value 100, and then prints the contents of **X** to the output. The same method would be used for printing a matrix of any dimension to the output.

```
PROC IML;
   x=100;
   PRINT x;
QUIT;
```

Output – Example 1.3:

```
                              x
                            100
```

By printing a matrix to the output the programmer can report and verify the exact contents of a matrix. The SHOW command, to be discussed later, can also be used to verify the properties of a matrix.

1.9 Creating a Matrix

IML can create a matrix in several different ways. The three most common ways to create a matrix in IML are as follows:

- Creating a matrix by manual entry
- Creating a matrix from a SAS data set
- Creating a matrix from a text file

Though data come in forms other than SAS data sets and text files, for infrequent or one-time use it is often easier to first convert the data from its native form into a SAS data set or a text file and then use one of the methods presented in this chapter for creating a matrix, rather than developing a new method for creating a matrix from the data in its native form.

1.9.1 Creating a Matrix Using Manual Entry

When using a piece of paper to create a matrix, a couple of brackets are drawn with numbers (referred to as matrix *elements*) in the middle in a "rows and columns" format. Creating a matrix by manual entry is similar. The following three examples demonstrate how to create the three types of matrices by entering their values as part of the code for a PROC IML step.

Example 1.4 – Creating a numeric matrix by manual entry:

This is an example of how to create a 5 × 3 numeric matrix, **A**, by manual entry in IML. The matrix values are all contained within braces ({ }). Each row in the matrix is separated by a comma (,). Each value within a row is separated by a space. The

```
PRINT a;
```

statement prints the matrix **A** to the output.

```
PROC IML;
   a={10 88 92,
       9 91 95,
      10 85 83,
       8 90 87,
      10 92 96};
   PRINT a;
QUIT;
```

Output – Example 1.4:

	a	
10	88	92
9	91	95
10	85	83
8	90	87
10	92	96

Example 1.5 – Creating a character matrix by manual entry:

This is an example of how to create a 5 × 2 character matrix, **POSITIONS**, by manual entry in IML. Again, notice that a space separates items within a row of the matrix. Notice also that case is not preserved in the output. Mixed case is used in the code, but all upper case is observed in the output. It can also be seen in the output for this example that although values in a character matrix are left-justified in the output, values in a numeric matrix (see previous example) are right-justified.

```
PROC IML;
   positions = {Rob     President,
                Joe     Supervisor,
                Erich Supervisor,
                Lisa     Manager,
                Mike     Director};
   PRINT positions;
QUIT;
```

Output – Example 1.5:

```
                    positions
        ROB            PRESIDENT
        JOE            SUPERVISOR
        ERICH          SUPERVISOR
        LISA           MANAGER
        MIKE           DIRECTOR
```

Example 1.6 – Variations of quotes and commas:

Commas and quotation marks (single or double) have different effects on the construction of matrices. Elements separated by a space are included in the same row but in different columns. If character strings are not contained within quotation marks they will appear in the output in all uppercase; however, if character strings are contained within quotation marks they will retain the case in which they are coded. Commas appearing within quotation marks do not separate rows as they do when not within quotation marks. Likewise, spaces appearing within quotation marks do not separate items within a row as they do when not within quotation marks. A mixture of single and double quotation marks can be used to make quotation marks appear in a matrix element; however, the reader should be cautioned that open quotations (the failure to end quoted text) can lead to complicated errors.

```
PROC IML;
   names1 = {'Rob' 'Joe' 'Erich' 'Lisa' 'Mike'};
   names2 = {Rob 'Joe' Erich 'Lisa' Mike};
   names3 = {'Rob', 'Joe', 'Erich', 'Lisa', 'Mike'};
   names4 = {'Rob, Joe', 'Erich Lisa', "O'Connor"};
   PRINT names1,names2,names3 names4;
QUIT;
```

Output – Example 1.6:

```
                    names1
        Rob    Joe    Erich Lisa   Mike
```

Each of the names in the **NAMES1** row vector is in a separate row as they were each separated by a space. Each of the names in the **NAMES1** row vector maintained its mixed case as it appeared in the code because each name was contained within quotation marks.

Output – Example 1.6 (continued):

```
                    names2
        ROB    Joe    ERICH Lisa   MIKE
```

The names "Joe" and "Lisa" maintained mixed case because they were contained within quotation marks within the code. The other three names appear in all uppercase because they were not contained within quotation marks.

Output – Example 1.6 (continued):

```
names3 names4

Rob     Rob,    Joe
Joe     Erich   Lisa
Erich   O'Connor
Lisa
Mike
```

Separating the elements in **NAMES3** with commas causes these elements to appear in separate rows, but in the same column. **NAMES4** demonstrates that character strings contained within quotes will be considered a single element and will be contained in the same row and column–even if a comma appears within the quotes. The mixture of single and double quotes in the coding of the name O'Connor in **NAMES4** allows the apostrophe to appropriately appear in the output.

It can also be noted that separating matrix labels in the PRINT statement with a space prints the two matrices on the same line of output if space allows. Separating matrix labels in the PRINT statement with a comma prints the second matrix below the first.

1.9.2 Creating a Matrix from a SAS Data Set

SAS data sets contain both rows and columns as with matrices. However, the format is different. The columns of a SAS data set represent the data variables. The rows of a SAS data set represent data records, also referred to as observations.

Because of the versatility of IML, it can be useful to manipulate data in the form of a matrix rather than in the form of a data set. Individual values in a matrix can be accessed by unique row and column indicators. Functions can also be applied to individual values, entire rows, entire columns, or subsections of a matrix. With data sets, it is fairly natural to apply a function across values in a column, but it is a little more complex to apply a function to an entire row as that would cross multiple variables that each have to be indexed and may include both numeric and non-numeric values. It is also helpful that IML does have the ability to handle large matrices (hundreds of rows and hundreds of columns). This functionality is not available with other software packages that may limit the number of rows and/or columns of matrices.

For the sake of not typing all the numbers, letters, or words in the matrix when that information is available in a data set, here are some examples of creating a matrix from an existing SAS data set:

Example 1.7 – Creating a matrix from a data set (1):

Previously, matrix **A** was defined to have the following five rows and three columns:

$$
\mathbf{A} = \begin{bmatrix} 10 & 88 & 92 \\ 9 & 91 & 95 \\ 10 & 85 & 83 \\ 8 & 90 & 87 \\ 10 & 92 & 96 \end{bmatrix}.
$$

These same five rows and three columns may exist as five records of a data set with three columns, as follows:

DATA Set **DSET_A**

_N_	A1	A2	A3
1	10	88	92
2	9	91	95
3	10	85	83
4	8	90	87
5	10	92	96

The following code will create the matrix **A** from the existing SAS data set **DSET_A**:

```
PROC IML;
    USE dset_a;
    READ ALL VAR {a1,a2,a3} INTO a;
    CLOSE dset_a;
    PRINT a;
QUIT;
```

The

```
USE dset_a;
```

statement identifies **DSET_A** as the name of the SAS data set that will be used by the IML procedure. It opens the SAS data set for access by the procedure. The READ ALL VAR statement tells IML which columns to include in the matrix and gives the matrix the name **A**. The ALL VAR option indicates that all variables listed within the brackets ({ }) are to be used in the creation of the resulting matrix—not that all variables in the open SAS data set are to be read into the resulting matrix unless the variable names for all the variables in the open SAS data set are specified within the brackets. Two other options for reading in variables from a SAS data set are to use either the _NUM_ or _ALL_ keywords. Replacing the READ statement with

```
READ ALL VAR _NUM_ INTO a;
```

results in all numeric variables from the open SAS data set being read into the matrix **A**, and replacing the READ statement with

```
READ ALL VAR _ALL_ INTO a;
```

results in all variables from the open SAS data set being read into the matrix **A**. The

```
CLOSE dset_a;
```

statement closes the data set **DSET_A**, that was opened for access with the

```
USE Dset_A;
```

statement. The CLOSE statement is not required for this example to work properly. However, it is good practice to close a data set after it has been used, to avoid possible conflicts. If another procedure attempts to access a SAS data set that is already open, being used by another process, an error message will display in the log indicating a conflict. The first process must finish using the SAS data set before another can start. Closing the data set using the CLOSE statement will ensure the data set is closed and can then be opened and accessed by another process. The PRINT statement prints the newly created matrix **A** to the output.

Output – Example 1.7:

	A	
10	88	92
9	91	95
10	85	83
8	90	87
10	92	96

This example is one of the more simple methods used to create a matrix from a SAS data set. The next example is a bit more complex.

Example 1.8 – Creating a matrix from a data set (2):

Consider the following SAS data set, **HOMEWORK**. It is proposed that a linear model be defined using the data associated with variables **X1** through **X4** as the matrix **X** and the data associated with the variable **Y** as the column vector **Y**. The linear model is subject to the condition that only data for "Problem 1" be included.

Data Set *Homework*

_N_	PROBLEM	Y	X1	X2	X3	X4
1	1	3	20	140	10	68
2	1	2	19	110	7	64
3	1	3	20	190	11	72
4	1	3	23	250	11	71
5	2	4	21	110	6.5	62

The following code can be used to create the matrix **X** from data columns **X1** through **X4** and the column vector **Y** from the data column **Y**, subject to the condition that only data for "Problem 1" be included:

```
PROC IML;
   USE homework;
   READ ALL VAR {y} INTO y WHERE (problem = 1);
   READ ALL VAR {x1,x2,x3,x4} INTO x WHERE (problem = 1);
   CLOSE homework;
   PRINT y x;
QUIT;
```

Here, two matrices, or more precisely a column vector and a matrix, are created from the same SAS data set with two different READ statements. The **Y** variable goes to the **Y** matrix and the **X** variables go to the **X** matrix. It can also be the case that multiple matrices can be defined by utilization of the same variable. The WHERE statement restricts the rows, included in the matrix **X** and the column vector **Y**, to include only the records that pertain to the variable **PROBLEM** with value equal to 1. Similarly, two matrices could have been created using data with the restriction that only records with **PROBLEM** = 2 be included.

The following is an illustration of what the resulting column vector **Y** and the resulting matrix **X** might look like if the data set **HOMEWORK** only contained the five observations previously listed:

Resulting Matrices **Y** and **X**

$$
Y = \begin{bmatrix} 3 \\ 2 \\ 3 \\ 3 \end{bmatrix} \quad X = \begin{bmatrix} 20 & 140 & 10 & 68 \\ 19 & 110 & 7 & 64 \\ 20 & 190 & 11 & 72 \\ 23 & 250 & 11 & 71 \end{bmatrix}
$$

Output – Example 1.8:

```
        Y          X
        3         20        140         10         68
        2         19        110          7         64
        3         20        190         11         72
        3         23        250         11         71
```

1.9.3 Creating a Matrix from a Text File

If the data to be read into a matrix is contained in a text file, the process for importing the data into a matrix is a little more complex. The complexities result from the different ways in which the original data may be formatted to appear in the text file. The user must inform IML as to how the data are stored in the text file and how they are to be read in. Creating a matrix from an external file can be a two-step process:

1. Read the data into a SAS data set.
2. Create a matrix from the SAS data set.

There are ways of bypassing the first step. The following four examples use different approaches to read in text from an external text file. These different methods illustrate differences in the format of the data in the text file and how the INPUT statement in PROC IML can be modified to read in the data according to their format. Accompanying the examples is a discussion of the different types of INPUT statements being illustrated.

Example 1.9 – Creating a matrix from a text file using list input:

The first example has a text file located at A:\CLASS DATA\CLASS A.TXT. Opening the text file in a text viewer it appears as follows:

```
    Class A Data     ←(first line of the file)
                     ←(single blank space)
     Student    Score
                     ←(single blank space)
     Rob          92
     Joe          95
     Erich        83
     Lisa         87
     Mike         96
```

List input can be used when all the data values are separated by a delimiter. By default, the delimiter is blank (space or tab) characters. In a SAS DATA step, the delimiter may be changed to a comma (,) as with .CSV files or to some other character or string of characters. However, the ability to change the delimiter from blank characters to some other character does not exist in PROC IML. If delimiters exist in the text file other than blanks, a second step can be added. The text file can first be read into a SAS data set using a DATA step and then the SAS data set can be used to create matrices using PROC IML as in the previous example. Data are read from left to right starting with the first character string prior to the delimiter (the first blank). That first character string, be it numeric or non-numeric, is considered the value for the first variable. After the blank character(s), the next non-blank character string is considered the value for the second variable. This continues to the end of the line. The text file A:\CLASS DATA\CLASS A.TXT is set up such that list input is an appropriate way to read in the data. Blank spaces separate each of the recorded values. The text in the first line of the data file is a title and is not intended as a data value. In the resulting data set and/or matrices, this line of text is to be omitted. The same is true of the blank lines and the third line containing the column headers. Though pertinent in a printout of the data file, these occurrences are not data values and consequently are not to be read in. The code must indicate the location of the data values that are to be read into the data sets and/or matrices.

The following code creates the matrix **CLASS_A** from the text file A:\CLASS DATA\CLASS A.TXT using list input:

```
PROC IML;
    FILENAME inclass 'a:\class data\class a.txt';
    INFILE inclass MISSOVER;
    INPUT ///;
    students=J(5,1,"Student");
    scores=J(5,1);
    DO i=1 TO 5;
        INPUT student $ score;
        students[i]=student;
        scores[i]=score;
    END;
    CLOSEFILE inclass;
    PRINT students scores;
QUIT;
```

The FILENAME statement indicates the location of the external text file and assigns that location the name INCLASS. The name INCLASS then becomes a pointer to the file. No new INCLASS file is created, taking up storage space. The file, A:\CLASS DATA\CLASS A.TXT, is just given a name whereby it can be subsequently referenced in the code without the user having to rewrite the entire file path and name each time it is referenced.

The INFILE statement also opens the external file in preparation for it to be read. The MISSOVER option indicates to IML how it is to handle missing values in the data. If, while reading across a record in the text file, there are no more non-blank character strings and yet there are still variables that have not been assigned values for that row, the MISSOVER option records each of the as yet unassigned values for that row to be missing. SAS then begins reading data for the next observation from the next line.

The first INPUT statement reads in no variables. The three back slashes, ///, indicate that IML should skip the first three lines of the text file containing the title, blank and column labels.

The

```
students=J(5,1,"Student");
```

and

```
scores=J(5,1);
```

statements create the 5 × 1 column vectors **STUDENTS** and **SCORES**. Both statements use the function J which has the effect of producing a matrix of any dimension and populating it with any value (the same value for every element). In the first case, The column vector **STUDENTS** is created with the J function. The first element of the function is 5, the number of rows. The second element of the function is 1, the number of columns. The third element of the function is "Student", the character string that will initially populate each element of the newly created 5 × 1 column vector **STUDENTS**. The 5 × 1 column vector **SCORES** is populated initially with the numeric value 1 in each element, the default when the third value of the J function is left unspecified. These two vectors will hereafter be populated with the actual data values. The **STUDENTS** vector is initialized with the word "Student," a character string of length seven. This allows for the elements of the vector **STUDENTS** to be character of maximum length seven. If there are character strings in the file that are of a length greater than seven, the vector element will truncate the character string to length seven and all subsequent characters will be dropped. So, the initial value should have enough character length to support the longest student name in the data. Because no value was specified for the elements of the **SCORES** column vector, each element will contain the numeric value 1. If desired, the elements could have been initialized to a different value by adding a third expression to the J function. For example, to initialize all elements of the **SCORES** column vector to the value 0, the statement could have read

```
scores=J(5,1,0); .
```

Initializing elements to a value that is unlikely to appear in the actual data can be beneficial. Once the complete program is run, the user can easily see which initial values were replaced with actual data and which were not.

The DO loop iterates five times. That is due to the fact that it is known that there are only five total lines of data values to read from the text file.

Having skipped four lines with the previous INPUT statement (three skipped by the three back slashes and the fourth skipped because the INPUT statement read the fourth line but created no variables because no variable names were included in that first INPUT statement), the INPUT statement within the DO loop gives a name to the variables **STUDENT** and **SCORE**. According to list input, the $ symbol follows all character variables. All variables in the INPUT statement not followed by the $ symbol are considered numeric. With each pass of the DO loop, the INPUT statement reads a line of data values. The

```
students[i]=student;
```

and

```
scores[i]=score;
```

statements read the current data values for the variables **STUDENT** and **SCORE** into the ith row of the column vectors **STUDENTS** and **SCORES**, respectively. The **I** is a counter variable that increments with each pass of the DO loop.

The

```
CLOSEFILE inclass; .
```

statement closes the text file. The PRINT statement prints the two column vectors, **STUDENTS** and **SCORES**, side-by-side.

Output – Example 1.9:

```
          students     scores
          Rob              92
          Joe              95
          Erich            83
          Lisa             87
          Mike             96
```

Example 1.10 – Creating a matrix from a data set (created from text file using list input):

In this example a SAS data set, CLASS_B, is first created from the text file and then the two column vectors, **STUDENTS** and **SCORES**, are created from a CLASS_B data set using PROC IML. Often times it is preferable to use the two-step approach of first creating a SAS data set from the text file and then creating matrices from the SAS data set as with this example rather than the one-step approach of creating matrices in PROC IML from the text file directly as with example 1.9. There are more options available in the DATA step for reading in data from different types of text files than there are in PROC IML. As well, certain types of data manipulation may be more easily performed in a DATA step than in PROC IML. The user will have to decide which method, the one-step approach, or the two-step approach is more appealing for a given situation.

Consider the following text file, CLASS B.TXT. Note that it is the same as CLASS A.TXT from the previous example except that the first three lines of CLASS A.TXT are not included in CLASS B.TXT.

File A:\CLASS DATA\CLASS B.TXT:

```
Rob        92
Joe        95
Erich      83
Lisa       87
Mike       96
```

In the following code, a SAS data set, CLASS_B, is created from the text file A:\CLASS DATA\CLASS B.TXT using list input in a DATA step. The SAS data set, CLASS_B, is then printed via PROC PRINT. Next, the SAS data set CLASS_B is used in PROC IML to create the two column vectors **STUDENTS** and **SCORES**:

```
DATA class_b;
    FILENAME inclass 'a:\class data\class b.txt';
    INFILE inclass MISSOVER;
    INPUT student $ score;
RUN;
PROC PRINT DATA=class_b;
RUN;
PROC IML;
    USE class_b;
    READ ALL VAR {student} INTO students;
    READ ALL VAR {score} INTO scores;
    CLOSE class_b;
    PRINT students scores;
QUIT;
```

The code involves two basic steps. The first is reading in data from a text file into a
SAS data set using a DATA step. The second step involves creating matrices from
SAS data sets. This procedure was covered in examples 1.7 and 1.8.

Output – Example 1.10:

Obs	student	score
1	Rob	92
2	Joe	95
3	Erich	83
4	Lisa	87
5	Mike	96

The above output contains the newly created data set CLASS_B and is generated by
the PROC PRINT step.

Output – Example 1.10 (continued):

students	scores
Rob	92
Joe	95
Erich	83
Lisa	87
Mike	96

The column vectors, **STUDENTS** and **SCORES**, from this example output result
from the IML procedure and are identical to those created in the previous example.

*Example 1.11 – Creating a matrix from an external text file (using list input and
a trailing @):*

This example includes a variation of the list input by including the trailing @ feature,
also known as the line hold control. The trailing @ feature is characterized by the @

character at the end of the INPUT statement. Other than the trailing @, the INPUT statement is the same as it is for other list input processes. The trailing @ allows for text to be read from a text file that has multiple records on each line. The line is read until no more non-blank characters are detected and then it proceeds to the next line. If the file is viewed in a text viewer it appears as follows:

```
Rob 92 Joe 95 Erich 83 Lisa 87
Mike 96
```

In the following code, a SAS data set, CLASS_C, is created from the text file C:\TEMP CLASS C.TXT using list input with a trailing @. Then the column vectors **STUDENTS** and **SCORES** are created. Both the data set and the two column vectors are created within the PROC IML step. In this case the code represents a two-step process as with the previous example, but both steps are performed within the PROC IML step. This demonstrates the fact that the user has the option of reading the data from a text file into a SAS data set from within PROC IML or within a DATA step. Once the data set is created within a PROC IML step, it must then be closed before it can be again opened to be accessed for the data from the data set to be read to create the two column vectors.

The INFILE statement contains the FLOWOVER option. This option tells IML to go to the next line when it is unable to find a record on its current line. If this option is not set in conjunction with the trailing @, PROC IML will remain on the first line of the file until the DO loop ends. In the current example, the DO loop is scheduled to end when the pointer reaches the end of the file. If the FLOWOVER option is not included, it would never reach the end of the file and would continue until the user intervened.

The following code can be used to create the matrix **CLASS_C**:

```
PROC IML;
    FILENAME inclass 'c:\temp\class c.txt';
    INFILE inclass FLOWOVER;
    student="student";
    CREATE class_c VAR {student score};
    DO DATA;
        INPUT student $ score @;
        APPEND;
    END;
    CLOSE class_c;
    CLOSEFILE inclass;
    USE class_c;
    READ ALL VAR {student} INTO students;
    READ ALL VAR {score} INTO scores;
    CLOSE class_c;
    PRINT students scores;
QUIT;
PROC PRINT DATA=class_c;
RUN;
```

The trailing @ tells IML to remain on the current line after it reads the current record to look for multiple records. If this option is not set, IML will only read the first record found on each line, thus losing data.

Output – Example 1.11:

```
                     students      scores
                     Rob              92
                     Joe              95
                     Erich            83
                     Lisa             87
                     Mike             96
```

The above output is generated from the PRINT statement in the PROC IML step. The below output contains the newly created data set CLASS_C and is generated by the PROC PRINT step.

Output – Example 1.11 (continued):

```
             Obs      student      score
              1         Rob          92
              2         Joe          95
              3         Erich        83
              4         Lisa         87
              5         Mike         96
```

Example 1.12 – Creating a matrix from a data set (created from text file using column input):

This example demonstrates reading in data from a text file using *column input*. Column input identifies different variables in a text file by the vertical columns in which they reside in the text file. Consequently, data in the text files must reside in vertical columns in order to be read in using column input. The benefit of column input over list input is that it does not rely on delimiter characters to distinguish the columns of data. Using list input can be problematic with large complex text files because legitimate data values may contain the delimiter character as one of the characters in the data value.

Consider the name "Doe, John" as the data value. It appears with both a comma (,) character and a blank space character (after the comma). Using list input the user can not use a comma or a blank space as the delimiter to read in the data. If the INPUT statement attempts to read in name as one data value and a delimiter causes it to be read in as two values, the variable name will get the first character string, the last name, and whichever variable is defined next will get the second character string, the first name. If the variable that was defined to follow the variable name is defined as numeric, SAS will report a syntax error when a non-numeric data value is read into a numeric variable. If the variable that was defined to follow the variable name is defined as character, SAS may not report a syntax error at all, and yet the data will be corrupted.

Using column input does require that the data in a text file be arranged in columns. The INPUT statement identifies the columns in which each variable resides by indicating the starting and ending location of each column of data. These columns in the text file need not have delimiters of any type, but must be lined up vertically. The text file C:\TEMP\CLASS D.TXT has the following form:

```
Bachler, Rob   76
Campbell, Joe 72
Hoelzer, Erich72
Clayton, Lisa 67
Smith, Mike    73
```

Note that there is no blank space character separating the name "Hoelzer, Erich" from the value 72. As previously mentioned, delimiters are not needed or even regarded when column input is employed. In fact with column input, columns or portions of columns may be read in multiple times within a single INPUT statement. As well, columns do not have to be read in a certain order.

In the following code, a SAS DATA step is used to first create the SAS data set CLASS_D from the text file C:\TEMP\CLASS D.TXT using column input. Second, the column vectors **STUDENTS** and **HEIGHTS** are created from the SAS data set CLASS_D within PROC IML. This two-step approach, the data step and then the PROC IML step, is employed as column input is not directly supported in PROC IML.

In the INPUT statement the starting and ending locations of the two columns are specified. The text file column containing the student data values begins at location 1 and ends at location 14. The dollar sign ($) specifies the **STUDENT** variable as character. The text file column containing the heights data values begins at location 15 and ends at location 16. The absence of the dollar sign specifies the heights variable as numeric. The PROC PRINT step prints the newly-created CLASS_D data set to the output. The PROC IML step then reads the two variables **STUDENT** and **HEIGHT** from the CLASS_D data set and uses them to create the column vectors **STUDENTS** and **HEIGHTS**. The two column vectors are then printed to the output. Note that although the variable name **STUDENT** differs from the vector name **STUDENTS** in this example for easy identification, it need not differ in practice.

```
DATA class_d;
    FILENAME inclass 'c:\temp\class d.txt';
    INFILE inclass;
    INPUT student $ 1-14 height 15-16;
RUN;
PROC PRINT DATA=class_d;
RUN;
PROC IML;
    USE class_d;
    READ ALL VAR {student} INTO students;
```

```
    READ ALL VAR {height} INTO heights;
    CLOSE class_d;
    PRINT students scores;
QUIT;
```

Output – Example 1.12:

Obs	Student	Height
1	Bachler, Rob	76
2	Campbell, Joe	72
3	Hoelzer, Erich	72
4	Clayton, Lisa	67
5	Smith, Mike	73

The above output contains the newly created data set CLASS_D and is generated by the PROC PRINT step.

Output – Example 1.12: (continued):

students	heights
Bachler, Rob	76
Campbell, Joe	72
Hoelzer, Erich	72
Clayton, Lisa	67
Smith, Mike	73

The above section of output contains the newly created column vectors **STUDENTS** and **HEIGHTS** and were created in the PROC IML step.

Example 1.13 – Creating a matrix from an external text file (using formatted input):

Formatted input identifies the starting point of each new data value as well as both the length to assign to the variable and the format of the variable. A date, for example may appear as a data value

March 19, 2009.

This data value is non-numeric due to the characters in the name of the month as well as the comma. The same date may also appear as

3/19/2009

or

19Mar2009

All of these designations represent the same date, but they appear in different formats. As such, it will be difficult for the user to work with dates in computer algorithms if those dates are considered character text. For example, if the values March 19, 2009 and June 25, 2009 are ordered as character text, the June date will appear prior to the March date simply because alphabetically, J (the first letter of "June") comes before M (the first letter of "March"). To order date values chronologically, the date values must be treated such that the format in which they appear is not a basis for the order in which they appear—regardless of format, the chronological order remains the same.

SAS has a special way of handling date variables. It is called a "SAS date." A SAS date is computed as the number of days after the date January 1, 1960. So, the SAS date for January 1, 1960 is zero. The SAS date for January 10, 1960 is 9. The SAS date for December 31, 1959 is -1, and so forth. Choosing this almost arbitrary date at a convenient point along the timeline allows all dates to be considered as numeric values. One can then order dates chronologically, subtract two dates to determine the duration between the two, etc.

In order for the raw text "March 19, 2009" to be read in as a SAS date, the user must include in the INPUT statement the format in which the date appears. The format of a raw data value read into SAS from an external file is referred to as an *informat*, and the way the data appear in SAS output is considered the data *format*. For example, the informat of "March 19, 2009" is ANYDTDTE14., and its associated SAS date is 17975, indicating that March 19, 2009 is 17,975 days past January 1, 1960. To read the raw text "March 19, 2009" into SAS appropriately, the ANYDTDTE14. informat would appear in the INPUT statement.

There are many other examples for which formatted input may be appropriate. These examples include reading in times, full names, dollar amounts, and phone numbers to name a few. Additionally, there may be raw data with a format that is not recognized by one of the specific informats currently supported by SAS. In that case, data may have to be read in as text and then converted, say, within a DATA step. For example, if a Social Security number would be read in as plain text. There is no special informat for identifying a Social Security number as it is read into SAS from an external text file. Though, after being read into SAS the text could then be manipulated in a DATA step if one wished to conduct an analysis involving the digits.

The objective of this example is to import data from a text file similar to that of the previous example. The text file in this example differs by having zeros as numeric delimiters about the numbers, it contains record numbers as the first three characters of each record, and a fourth variable has also been added to the data containing the date of birth of the individual. In a text viewer, the text file C:\TEMP\CLASS E.TXT appears as follows:

```
001Rob   0920001/21/1969
002Joe   0950003/14/1971
003Erich0830005/15/1972
004Lisa  0870009/06/1971
005Mike  0960011/12/1965
```

where the zeros are used as placeholders that represent text that is not a part of the data you want to include.

The following code reads in the text file C:\TEMP\CLASS E.TXT and creates the SAS data set CLASS_E and the column vectors **STUDENT**, **SCORE**, and **DOB** (representing the date of birth of the individuals). The data are first read into the SAS data set CLASS_E. The

```
student="xxxxx";
```

statement initializes the **STUDENT** variable, indicating that it is going to contain a character string and allocates 5 characters length for the variable. Including fewer "x" values in the initialization will lead to data values in the **STUDENT** variable being truncated. The CREATE statement creates the data set CLASS_E and indicates the variables it will contain. The DO DATA loop steps through each record of the text file, importing data values into the appropriate variables in the resulting CLASS_E data set. The INPUT statement includes pointer controls (@4 and @9, and @14) to indicate the exact starting location of each data value for a given row in the text file. The $5. indicates the variable **STUDENT** is a text string of length five. The 3. indicates the variable **SCORE** is a number of length three. The MMDDYY10. indicates the variable **DOB** is a date of length ten. By specifying these exact variable types and lengths, none of the extra zeros are included and **DOB** can be read in as a SAS date.

```
PROC IML;
    FILENAME inclass 'c:\temp\class e.txt';
    INFILE inclass;
    student="xxxxx";
    CREATE class_e VAR {student score dob};
    DO DATA;
        INPUT @4 student $5. @9 score 3. @14 dob MMDDYY10.;
        APPEND;
    END;
    CLOSE class_e;
    CLOSEFILE inclass;
    USE class_e;
    READ ALL VAR {student} INTO students;
    READ ALL VAR {score} INTO scores;
    READ ALL VAR {dob} INTO dob;
    CLOSE class_e;
    PRINT students scores dob;
QUIT;
PROC PRINT DATA=class_e;
    FORMAT dob DATE9.;
RUN;
```

Output – Example 1.13:

students	scores	DOB
Rob	92	3308
Joe	95	4090
Erich	83	4518
Lisa	87	4266
Mike	96	2142

The above output contains the newly created column vectors **STUDENTS, SCORES,** and **DOB**. Though the data values under **DOB** appear in SAS date format for illustration, they could also have been formatted to have a more common date appearance by changing the PRINT statement to read

```
PRINT students scores dob [FORMAT=DATE9.];.
```

Output – Example 1.13 (continued):

Obs	STUDENT	SCORE	DOB
1	Rob	92	21JAN1969
2	Joe	95	14MAR1971
3	Erich	83	15MAY1972
4	Lisa	87	06SEP1971
5	Mike	96	12NOV1965

In addition to formatted input, list input also can take advantage of informats. Consider the following example:

Example 1.14 – Informats in list input:

Consider the text file C:\TEMP\CLASS F.TXT, that includes the variables **NAME, SCORE,** and **DOB**. In a text viewer, the text file appears as follows:

```
Rob 092 01/21/1969
Joe 095 03/14/1971
Erich 083 05/15/1972
Lisa 087 09/06/1971
Mike 096 11/12/1965
```

Neither column input nor formatted input would properly read in this file due to the overlap in the starting and ending points of the potential columns of data for the different records. List input must be used with an informat to read the **DOB** variable correctly.

The following code can be used to properly read in the data. The program includes both list input and informats.

```
PROC IML;
    FILENAME inclass 'c:\temp\class f.txt';
    INFILE inclass;
    student="xxxxx";
    CREATE class_f VAR {student score dob};
    DO DATA;
        INPUT student $ score dob MMDDYY10.;
        APPEND;
    END;
    CLOSE class_f;
    CLOSEFILE inclass;
    USE class_f;
    READ ALL VAR {student} INTO students;
    READ ALL VAR {score} INTO scores;
    READ ALL VAR {dob} INTO dob;
    CLOSE class_f;
    PRINT students scores dob;
QUIT;
PROC PRINT DATA=class_f;
    FORMAT dob DATE9.;
RUN;
```

The code differs slightly from the code in the previous example. In the INPUT statement, a dollar sign follows the character variable name **STUDENT**. No informat
follows the numeric variable name **SCORE**. The MMDDYY10. format follows the
dob variable name. By not specifying a specific starting point for the variables, list
input specifies that a blank space delimiter will indicate the separation between one
variable value and the next. A length is also not specified for the **STUDENT** and
SCORE variables as doing so would override the list input use of the blank space
delimiter. It is very important to note that in the text file C:\TEMP\CLASS F.TXT
each of the values in a given record are separated by a single space. This is key. If
list input is used with strictly character and numeric formats, values in a record can
be separated by one space or by a variable number of spaces. However, if informats
are used, values in the text file should be separated by a single space only and not
by multiple spaces. SAS reads the record under the assumption that a single space
separates the values in the text file. If that is not the case, values will be read in
incorrectly and an invalid data error will result.

Output—Example 1.14:

students	scores	DOB
Rob	92	3308
Joe	95	4090
Erich	83	4518
Lisa	87	4266

```
              Mike                 96        2142

        Obs      STUDENT      SCORE            DOB
         1         Rob          92       21JAN1969
         2         Joe          95       14MAR1971
         3         Erich        83       15MAY1972
         4         Lisa         87       06SEP1971
         5         Mike         96       12NOV1965
```

The first table in the output results from the PRINT statement in the PROC IML step. The second table results from the PROC PRINT step.

1.10 Saving and Retrieving a Matrix

IML has a memory feature that allows the user to save matrices to a specified file, and then to retrieve those matrices from the file. This allows the user to create matrices within a PROC IML step, leave the IML programming environment by exiting the PROC IML step, perform other tasks in the SAS programming environment, and then in a second PROC IML step return to the IML programming environment, and retrieve the previously created matrices for further computations. Saving an IML matrix is analogous to creating a permanent SAS data set. In both cases formatted data are saved for future use. The following example demonstrates the use of some of the matrix storage functions of IML.

Example 1.15 – Saving a matrix:

This example creates a matrix, **A**, from the data set DATA_A (Example 1.7), saves the matrix **A** to storage, and erases the matrix **A** from local memory. The LIBNAME statement creates a permanent library MATRICES, by establishing a pointer to the location C:\SAS92\MATRICES. This sets up a location to which the matrix will be saved. The permanent library must be defined prior to being accessed within the PROC IML step. The USE, READ, and CLOSE statements read the data from the SAS data set DATA_A to create and then close the matrix **A**. The

```
RESET STORAGE = "Matrices";
```

command identifies the permanent library MATRICES as a location for the storage of matrices and modules. The

```
STORE a;
```

command stores the matrix **A** in the storage location, previously assigned as the MATRICES permanent library. The

```
FREE a;
```

command frees the matrix **A** of its values and increases the available local memory—essentially, the matrix **A** is deleted from local memory. Any attempt to access the matrix **A** from local memory will yield an error message indicating that the matrix **A** does not exist.

```
LIBNAME matrices "c:\sas92\matrices";
PROC IML;
    USE data_a;
    READ ALL VAR {a1, a2, a3} INTO a;
    CLOSE data_a;
    RESET STORAGE = "Matrices";
    STORE a;
    FREE a;
QUIT;
```

The next set of statements can then be used to load the matrix **A** into local memory so that it can be referenced within the current PROC IML step for further use. Again, the

```
RESET STORAGE = "Matrices";
```

command identifies the permanent library MATRICES as a storage location for matrices and modules. The SHOW statement with the NAMES option lists all matrices in the current IML memory. The first SHOW statement is intended to show that there are no matrices (i.e.- the matrix **A**) in local memory prior to the LOAD statement. The

```
LOAD a;
```

command loads the matrix **A** from its location in the storage library, "Matrices," into the current IML memory for use. The second SHOW statement is intended to show that the matrix A has been loaded into local memory and is available for use within PROC IML. The

```
PRINT a;
```

statement then demonstrates the use of the matrix **A** within PROC IML by printing it to the output.

```
PROC IML;
    RESET STORAGE = "Matrices";
    SHOW NAMES;
    LOAD a;
    SHOW NAMES;
    PRINT a;
QUIT;
```

Output—Example 1.15:

```
SYMBOL   ROWS   COLS TYPE   SIZE
------   ------ ------ ---- ------
Number of symbols = 0
(includes those without values)
```

The above output results from the first

```
SHOW NAMES;
```

statement and shows that in fact there are no matrices (i.e.- matrix **A**) in local memory prior to the LOAD statement.

Output—Example 1.15 (continued):

```
SYMBOL    ROWS    COLS TYPE    SIZE
------  ------  ------ ----  ------

A             5       3 num       8
Number of symbols = 1
(includes those without values)
```

The second output (above) is a result of the second

```
SHOW NAMES;
```

statement and confirms that the numeric matrix **A** has successfully been loaded into the local memory (as a result of the LOAD statement). The output indicates the name of the Matrix **A**, the number of rows, 5, and columns, 3, and that the matrix is a numeric matrix. It also indicates that the number of character spaces allocated to each item in the matrix (SIZE) is 8.

Output—Example 1.15 (continued):

```
                              a
            10           88           92
             9           91           95
            10           85           83
             8           90           87
            10           92           96
```

The third output is a result of the PRINT statement and demonstrates that the matrix **A** can be used within the current PROC IML step after it has been loaded into the local memory.

1.11 Chapter Exercises

1.1 Describe, compare, and contrast SAS matrices and SAS data sets. Include in your description a brief list of benefits associated with each.

1.2 Describe, compare, and contrast numeric SAS matrices and character SAS matrices. Include in your description a brief list of benefits associated with each (there may be some overlap in the answers to this and the previous exercise).

1.3 Explain why mixed matrices, matrices containing both numeric and character data values, are not typically as useful as strictly numeric and strictly character SAS matrices.

1.4 Give an example of a data file containing at least two numeric and two character variables. Identify each variable by name and type (numeric or character). Explain the advantages and disadvantages to analyzing these data in the form of a SAS data set vs. an IML matrix.

1.5 Write a program in IML that demonstrates how matrix algebra can be used to solve the following set of linear equations for the values a, b, and c:

$$5a + 3b = 11$$
$$6a + 4b - 3c = 5$$
$$4a + 1b - 4c = -6$$

1.6 Submit the following code to create the text file MATH TEST.TXT (you may choose an alternate file location). Use *list input* to create a 5×1 column vector, **SUBJECT**, using the first column of the text file; and a 5×2 matrix **SCORES**, using the second and third columns of the text file. Print the column vector and matrix to the output to verify they are correct.

```
DATA _NULL_;
INPUT subject $ pretest posttest;
FILE 'c:\temp\math test.txt';
IF (_N_=1) THEN PUT "Standardized Math Exam"//
"Subject Pretest Posttest";
PUT @1 subject @12 pretest +6 posttest;
DATALINES;
Adam 78 85
Aaron 86 89
Amy 84 85
Alison 77 88
Andrea 98 99
;
RUN;
```

1.7 Submit the following code to create the text file MATH TEST.TXT (you may choose an alternate file location). Use *column input* to create a 5×1 column vector, **SUBJECT**, using the first column of the text file; and a 5×2 matrix **SCORES**, using the second and third columns of the text file. Print the column vector and matrix to the output to verify they are correct.

```
DATA _NULL_;
INPUT subject $ pretest posttest;
FILE 'c:\temp\math test.txt';
IF (_N_=1) THEN PUT "Standardized Math Exam"//
"Subject Pretest Posttest";
PUT @1 subject @12 pretest +6 posttest;
DATALINES;
Adam 78 85
Aaron 86 89
```

```
Amy 84 85
Alison 77 88
Andrea 98 99
;
RUN;
```

1.8 Submit the following code to create the text file HEART RATE.TXT (you may choose an alternate file location). Use *formatted input* to create a 5×1 column vector, **SUBJECT**, using the first column of the text file; a 5×1 column vector, **DOB**, using the second column of the text file; and a 5×1 column vector, **HR**, using the third column of the text file. The dob column should be read in using SAS date informat. Print the column vectors to the output to verify they are correct. Use a SAS word date format to format the dates in the **DOB** vector. For example, the date in the first row will appear in the output as December 23, 1977.

```
DATA _NULL_;
INPUT subject $ dob $10. hr;
FILE 'c:\temp\heart rate.txt';
IF (_N_=1) THEN PUT "Stress-induced Heart Rate"//
"Subject     Birth      HR";
PUT @1 subject @9 dob +2 hr;
DATALINES;
Bobby 12/23/1977 68
Billy 04/09/1972 72
Betty 05/15/1975 65
Barry 02/12/1968 88
Bonnie 07/11/1987 75;
RUN;
```

1.9 Give an example of a specific use of the storage feature in IML for storing and retrieving matrices.

Chapter 2
IML Language Structure

READ, UPDATE, RESET, PRINT? Many commands, functions, calls, and operators are pretty easy to figure out without explanation — It might be expected that the READ statement read values into a matrix, the NCOL function identify the number of columns in a given matrix, and the + operator add elements in a matrix. However, some of the commands, functions, calls, and operators used in IML are less obvious. This section will help the reader understand the basic use of commands, functions, calls, and operators with some examples.

2.1 Statements

The IML language is primarily classified into statements and operators. The different types of statements include functions, calls, and commands and control statements. The purpose of this section is to include a few examples of functions, calls, commands, control statements, calls, and operators to get the reader started. The reader can then apply the knowledge learned to other IML statements.

2.1.1 Functions

Functions are automated procedures that use user-defined arguments for calculations or manipulations and return a result. The general form of a function is

$$result = FUNCTION\ (arguments);$$

where *arguments* includes matrix names, numbers, character strings, and expressions. SAS/IML has many functions to increase program efficiency by decreasing the amount of code used for certain calculations or manipulations.

If there is a function not available in PROC IML that would be beneficial the researcher can define modules, that work as functions, that can then be used in the same manner as those already defined in SAS/IML. The following examples include some of the more basic IML functions useful in the analysis of linear models:

J.J. Perrett, *A SAS/IML Companion for Linear Models*, Statistics and Computing, 33
DOI 10.1007/978-1-4419-5557-9_2, © Springer Science+Business Media, LLC 2010

Example 2.1 – The NROW and NCOL functions:

The NROW and NCOL functions compute the number of rows and the number of columns in a matrix, respectively.

```
PROC IML;
    names = {Lucas LeAnna , Jason Mollie , Brent Robert};
    rows = NROW(names);
    columns = NCOL(names);
    PRINT names rows columns;
QUIT;
```

Output – Example 2.1:

		ROWS	COLUMNS
NAMES			
LUCAS	LEANNA	3	2
JASON	MOLLIE		
BRENT	ROBERT		

Example 2.2 – The LOC function:

From its main description, "finds the nonzero elements of a matrix," one might question the LOC function's usefulness. How often is it useful to identify the nonzero elements of a matrix? The LOC function is useful for that, but is also has other seemingly unrelated uses. One such use is the avoidance of unnecessary looping. The following code uses two DO loops to find and print the odd numbers in each column of the matrix **M**.

```
PROC IML;
    m={1 2 3,4 5 7};
    DO i=1 TO NCOL(m);
        DO j=1 TO NROW(m);
            odd=MOD(m[j,i],2)>0;
            IF (odd)>0 THEN PRINT i (m[j,i]);
        END;
    END;
QUIT;
```

The MOD function computes the modulus, the remainder to the division of the value of element **M[j,i]** being divided by 2. If the result of the MOD function is 0, the value of element **M[j,i]** divides evenly into 2, and is consequently an even number. If it results in a number greater than zero, **M[j,i]** does not evenly divide into two and is consequently an odd number. The function

```
odd=MOD(m[j,i],2)>0;
```

is a logical statement resulting in **ODD** having the value of 1 if MOD(m[j,i],2)>0 is true (i.e.- the value of element **M[j, i]** is odd) and 0 if it is false. The value of **I** represents the column number of the element **M[j, i]** currently being analyzed. At each step the program prints the value of **I** and the value of **M[j, i]** to the output if **M[j, i]** does in fact contain an odd number.

Output – Example 2.2:

```
                              I
                              1                    1
                              I
                              2                    5
                              I
                              3                    3
                              I
                              3                    7
```

This next set of code uses the LOC function to replace the inner DO loop.

```
PROC IML;
   m={1 2 3,4 5 7};
   DO i=1 TO NCOL(m);
      odd=LOC(MOD(m[,i],2)>0);
      IF NCOL(odd)>0 THEN PRINT i (m[odd,i]);
   END;
QUIT;
```

Notice that in this case, reference is made to **M[,i]**. When the row of interest is not specified within the brackets, all rows are included in the reference. Consequently, the LOC function has the effect of checking all rows in **M[,i]** for whether the number in each element **M[j, i]** is odd or even. The result of

```
odd=LOC(MOD(m[,i],2)>0);
```

in this case is actually a column vector. Each element of the column vector **ODD** is the column number of the element **M[j, i]** found to be an odd number. After each iteration of the DO loop, the column reference, **I**, is printed along with all the numbers in that column found to be odd, **ODD**.

Output – Example 2.2 (continued):

```
                                           I
                                           1              1
                                           I
                                           2              5
                                           I
                                           3              3
                                                          7
```

Not only does the LOC function make the code more efficient, but it also makes it faster.

2.1.2 Calls

A "call" calls a subroutine or function. A difference between *calls* and *functions* is that functions generally return a single result. Calls are subprograms. A call may take one argument and create several new arguments in return. A good example of this is when computing eigenvalues and eigenvectors. A call may also take various arguments and perform an entire analysis. A good example of this is any one of the nonlinear optimization calls. Using the RENAME call renames a SAS data set, not computing anything at all!

Example 2.3 – Functions vs. calls (eigenvalues and eigenvectors):

This example computes eigenvalues and eigenvectors using functions, and then computes the same eigenvalues and eigenvectors using a call statement.

```
PROC IML;
   mat = {1 2 3,2 4 5,3 5 6};
   eigval1=EIGVAL(mat);
   eigvec1=EIGVEC(mat);
   CALL EIGEN(eigval2,eigvec2,mat);
   PRINT eigval1 eigval2,,eigvec1 eigvec2;
QUIT;
```

Output – Example 2.3:

```
                    EIGVAL1    EIGVAL2
                 11.344814 11.344814
                  0.1709152 0.1709152
                 -0.515729 -0.515729
```

```
         EIGVEC1                              EIGVEC2
    0.3279853   0.591009    0.7369762 0.3279853   0.591009   0.7369762
    0.591009  -0.736976    0.3279853 0.591009   -0.736976   0.3279853
    0.7369762   0.3279853  -0.591009  0.7369762   0.3279853 -0.591009
```

This example demonstrates two ways to compute eigenvalues and eigenvectors. The first way uses functions and the second way uses a call statement. Both methods produce the same results. However, the EIGEN call was set up to provide both eigenvalues and eigenvectors with one statement, thus eliminating a line of code and making a program that requires both eigenvalues and eigenvectors more efficient.

Call statements generally have the following form:

CALL name <(arguments)> ;

If a user-defined subroutine is created with the same name as an IML built-in subroutine, using a CALL statement will implement the IML built-in subroutine, whereas using a RUN statement will implement the user-defined subroutine. The following are examples of call statements.

Example 2.4 – RENAME Call:

The RENAME call renames a SAS data set. The following code sets the default library to TEMP and then changes the name of the data set OLD to NEW.

```
PROC IML;
   RESET DEFLIB=temp;
   CALL RENAME(old,new);
QUIT;
```

The following code would produce identical results:

```
PROC IML;
   CALL RENAME(temp,old,new);
QUIT;
```

2.1.3 Commands

Commands are used to perform specific system actions, such as storing and loading matrices and modules, or performing special data processing requests. The following is a sample list of some IML commands and the actions they perform.

Command	Action
APPEND	adds observations to the end of a SAS data set
CLOSE	closes a SAS data set
CREATE	creates and opens a new SAS data set for input and output
DELETE	marks observations for deletion in a SAS data set
EDIT	opens an existing SAS data set for input and output
FIND	finds observations
INDEX	indexes variables in a SAS data set
LIST	lists observations
PURGE	purges all deleted observations from a SAS data set
READ	reads observations into IML variables
REPLACE	writes observations back into a SAS data set
RESET DEFLIB	names default libname
SAVE	saves changes and reopens a SAS data set
SETIN	elects an open SAS data set for input
SETOUT	elects an open SAS data set for output
SHOW CONTENTS	shows contents of the current input SAS data set
SHOW DATASETS	shows SAS data sets currently open
SORT	sorts a SAS data set
SUMMARY	produces summary statistics for numeric variables
USE	opens an existing SAS data set for input

These IML commands can be used to control displayed output, get system information, manage SAS data sets, etc.

If running short on available space, the researcher can use commands to store matrices in the storage library, free the matrices of their values, and load them back later when they are again needed.

The following example illustrates how several IML commands are used. It makes use of the IML commands STORE, SHOW STORAGE, LOAD, USE, READ, EDIT, and REPLACE.

Example 2.5 – Updating values in a data set:

Often tables must be used in calculations to comply with regulations. In this example a table of depreciation rates is used to calculate the current value of a product based on the type and age of product. The SAS data set INFO contains the ID, type, and age of the various products. The data set also includes the variable **VALUE** with missing values currently recorded for each observation. This example would also work if the variable, **VALUE**, contained numbers. The matrix **DEP_TABLE** is a 5×15 matrix of regulated depreciation values dependent upon the type and age of a product.

```
DATA info;
   INPUT id $ type age value;
   DATALINES;
   0010132 3 7 .
   0010133 2 8 .
```

```
    0010134 4 2 .
    0010135 4 1 .
    0010136 3 4 .
RUN;

PROC PRINT DATA=info;
RUN;

PROC IML;
   dep_table = {15 13 10  8  6 3 1 1 1,
                17 14 11  7  5 3 1 1 1,
                20 17 13 11 10 6 4 1 1,
                25 20 15 12  8 5 3 1 1,
                30 25 20 15 10 5 1 1 1};
   STORE dep_table;
   SHOW STORAGE;
QUIT;

PROC IML;
   LOAD dep_table;
   USE info;
   READ ALL VAR{type,age} INTO info2;
   EDIT info;
   DO i=1 TO NROW(info2);
      value = dep_table[info2[i],info2[i,2]];
      REPLACE POINT i;
   END;
   CLOSE info;
QUIT;

PROC PRINT DATA = info;
RUN;
```

The DATA step creates the SAS data set INFO with values for the variables **ID**, **TYPE**, and **AGE**. The values for the variable **VALUE** are all defined as missing. The purpose of this program is to populate the variable VALUE with values from the depreciation table. The first PROC IML step creates a 5×9 matrix **DEP_TABLE** and stores that matrix in the IML storage library. When the first PROC IML step quits, all matrices, in this case **DEP_TABLE**, are lost from local memory. However, because **DEP_TABLE** was stored, it can later be retrieved from IML storage library and loaded back into local memory. The second PROC IML step loads the matrix **DEP_TABLE** back into local memory. Next, the variables **TYPE** and **AGE** are read into the matrix **INFO2** from the SAS data set INFO. The

```
    EDIT info;
```

statement opens the SAS data set INFO for editing. The DO loop iterates through each row of the matrix **INFO2**. For each row,

```
value = dep_table[info2[i],info2[i,2]];
```

sets the new value of the variable **VALUE** based on a value from the depreciation table contained in the matrix **DEP_TABLE**. The row to be referenced in
DEP_TABLE is based on info2 [i]. When only one reference is given within
the brackets, it is a row reference. Consequently, the *entire* row **I** of the matrix
INFO2 is being referenced. The column to be referenced in **DEP_TABLE** is based
on info2 [i,2]. Once the appropriate value from **DEP_TABLE** is identified
and assigned to the variable **VALUE**, The

```
REPLACE POINT i;
```

statement replaces the missing values in the data set INFO with the value obtained
from the table, **DEP_TABLE**. This occurs for each iteration of the DO loop, or in
other words, for each observation in the data set INFO. The POINT option allows the
replacement to apply to a given record number, matrix, etc. In this case, POINT i
indicates that values are to be updated for each of the **I** observations in the data set.

Output – Example 2.5:

Obs	ID	type	age	value
1	0010132	3	7	.
2	0010133	2	8	.
3	0010134	4	2	.
4	0010135	4	1	.
5	0010136	3	4	.

The first output of this code is a printout of the original SAS data set, INFO. This
results from the PROC PRINT step.

The variable **VALUE** has missing values for all observations. These values are
to be updated by the depreciation table.

Output – Example 2.5 (continued):

```
Contents of storage library = WORK.IMLSTOR Matrices:
DEP_TABLE
Modules:
```

The second output is a result of the

```
SHOW STORAGE;
```

command. This command instructs SAS to list the current contents of the IML storage library. The matrix of depreciation rates, DEP_TABLE, is the only item contained in the storage. This item was stored using the

```
STORE dep_table;
```

command.

Output – Example 2.5 (continued):

Obs	ID	type	age	value
1	0010132	3	7	4
2	0010133	2	8	1
3	0010134	4	2	20
4	0010135	4	1	25
5	0010136	3	4	11

The third output is the updated data set INFO with the depreciation rates updated according to the table.

2.1.4 Control Statements

SAS/IML has statements designed to control the flow of execution of statements in the program. These statements, such as DO-END, IF-THEN, and GOTO, can execute various sections of the program, and can do so multiple times.

The IF-THEN statement conditions the execution of a programming statement on specified criteria:

IF <specifies the condition to be satisfied>
THEN <executes a programming statement only if the condition in the IF statement was satisfied>

For example, the statements

```
IF score > 90
THEN grade = "A";
```

can be used to assign the grade of "A" for a score that is greater than 90. The IF-THEN statement conditions the assignment of "A" to **GRADE** on the value of **SCORE** being greater than 90.

The IF-THEN statement can be followed up by an ELSE statement to execute a programming statement when the IF condition is not satisfied. For example, the statements

```
IF score > 55
THEN status = "PASS";
ELSE status = "FAIL";
```

can be used to assign the status of "PASS" for a score that is greater than 55 and the status of "FAIL" if the score is not greater than 55. So, the IF-THEN statement conditions the assignment of "PASS" to **STATUS** on the value of **SCORE** being greater than 55. The ELSE statement conditions the assignment of "FAIL" to **STATUS** on the IF condition not being satisfied (**SCORE** not being greater than 55).

When there are more than one statement to be executed when the IF condition is met or, if an ELSE statement is included, the ELSE condition being met, then a DO-END section can be added. For example, the statements

```
IF score > 90
THEN DO;
   status = "PASS";
   grade = "A";
END;
```

can be used to both assign the status of "PASS" and the grade of "A" for a score that is greater than 90. The IF-THEN statement conditions the statements contained within the DO-END section to execute only when the IF condition is satisfied (the score is greater than 90). The number of statements contained within a DO-END section is not limited, but without the DO-END section, the IF-THEN statement can execute only one statement.

Another type of DO-END section, called DO-iterate, can also be used in programming. The DO-iterate statement is of the form

```
DO i = 1 TO 10 BY 2;
```

Where **I** is a variable name that is assigned by the user, 1 is a starting value assigned by the user, 10 is an ending value assigned by the user, and BY 2 is an optional specification of increment that is assigned by the user. Consequently, the following are examples of DO-iterate statements:

```
DO j = 5 TO 10;
DO k = 1 TO 100 BY 5;
```

Additionally, the sequence can be descending as with the following statement:

```
DO m = 100 TO 1 BY DESCENDING 2;
```

A DO-iterate statement starts a DO-END section that can execute the statements contained within that section multiple times prior to continuing with the execution of the statements that follow the DO-END section. For example, the following the statements

```
DO i = 1 TO 10;
   square = i*i;
   PRINT i square;
END;
```

iterate the statements within the DO-END section 10 times. The ten iterations make the following assignments

Iteration 1: **I**=1, **SQUARE**=1
Iteration 2: **I**=2, **SQUARE**=4
Iteration 3: **I**=3, **SQUARE**=9
Iteration 4: **I**=4, **SQUARE**=16
Iteration 5: **I**=5, **SQUARE**=25
Iteration 6: **I**=6, **SQUARE**=36
Iteration 7: **I**=7, **SQUARE**=49
Iteration 8: **I**=8, **SQUARE**=64
Iteration 9: **I**=9, **SQUARE**=81
Iteration 10: **I**=10, **SQUARE**=100

During each iteration, the PRINT statement sends the current values of **I** and **SQUARE** to the output. The DO-iterate statement is useful for stepping through the rows or columns of a matrix and for conducting simulations.

Example 2.6 – Control statements:

This example uses control statements to determine pass/fail grades based on exam scores.

```
PROC IML;
   names = {Albert,Bob,Alice,Christopher,Adam,David};
   scores = {50,86,89,74,45,88};
   grades=J(NROW(names),1,"xxxx");
   DO i=1 TO NROW(names);
      IF scores[i]<55 THEN
         grades[i] = "FAIL";
      ELSE grades[i] = "PASS";
   END;
   PRINT 'Names, Scores, and Grades for Exam 1';
   PRINT '(Criteria for Passing: 55+)';
   PRINT names scores grades;
QUIT;
```

Output – Example 2.6:

```
Names, Scores, and Grades for Exam 1
    (Criteria for Passing: 55+)
    NAMES              SCORES GRADES
    ALBERT                 50 FAIL
    BOB        .           86 PASS
    ALICE                  89 PASS
    CHRISTOPHER            74 PASS
    ADAM                   45 FAIL
    DAVID                  88 PASS
```

In this example, the DO-iterate statement indicates that the statements listed within the DO-END section will iterate a number of times equal to the number of rows in the **NAMES** column vector. The IF-THEN statement assigns the label "FAIL" to the column vector **GRADES** for all records that satisfy the condition **SCORE** < 55. All records that do not satisfy that condition are assigned the label "PASS" to the column vector **GRADES**.

2.2 Operators

2.2.1 Subscript Reduction Operators (Mathematical Operator)

IML provides several operators that can be used to perform basic mathematical operations on elements of subsections of matrices. The effect of these operators is the reduction of matrices to smaller dimensions. For this reason they are also referred to as *subscript reduction operators*. These operators include the following:

Operator	Action
+	addition
#	multiplication
<>	maximum
><	minimum
<:>	index of minimum
>:<	index of maximum
:	mean
##	sum of squares

Example 2.7 – Subscript reduction operators:

The following code demonstrates uses of subscript reduction operators on the matrix **A**.

```
PROC IML;
   a={1 2 3 4,
      5 6 7 8,
9 0 1 2};
   col_sum=a [+,];
   col_prod=a [#,];
   row_max=a [,<>];
   overall_min=a [><];
   row_indx_max=a [,<:>];
   row_indx_min=a [,>:<];
   col_1_3_mean=a [:,{1 3}];
   col_ssq=a[##,];
   mean_col_max=a[<>,][,:];
   PRINT a, col_sum, col_prod, row_max, overall_min,
         row_indx_max, row_indx_min, col_1_3_mean, col_ssq,
         mean_col_max;
QUIT;
```

Specifying a specific column or set of columns determines which column or set of columns the operator provide results for. By omitting the column reference, the operator provides results for all columns. The analog is true for rows as well. By omitting both row references and column references, the operator provides results for all matrix elements.

The statement

```
col_sum=a[+,];
```

omits the column reference and so will provide results for all columns. The summation operator in the row reference indicates that elements will be summed across the rows to provide column totals. The statement

```
col_prod=a[#,];
```

computes the product across the rows for each column, the column products. The statement

```
row_max=a[,<>];
```

omits the row reference and so will provide results for all rows. The maximum operator in the column reference indicates that the elements across the columns will be searched for the maximum value in each row, the row maximums. Because the statement

```
overall_min=a[><];
```

omits both row and column references, the operation will involve all elements of the matrix **A**. The statement identifies the minimum value of all the elements in the matrix **A**, the matrix minimum. The statement

```
row_indx_max=a[,<:>];
```

identifies the column index for each row maximum. The statement

```
row_indx_min=a[,>:<];
```

identifies the column index for each row minimum. The statement

```
col_1_3_mean=a[:,{1 3}];
```

specifies that results are to be provided for columns 1 and 3. The mean operator computes the mean of the elements of all rows for each of columns 1 and 3, the column means. The statement

```
col_ssq=a[##,];
```

computes the sum of the square of elements of all rows for each column. The statement

```
mean_col_max=a[<>,][,:];
```

includes two subscript reduction operators. The first operator produces a row vector containing the column maximums. The second operator computes the row means. Together, the two operators produce the mean of the column maximums.

Output – Example 2.7:

	a		
1	2	3	4
5	6	7	8
9	0	1	2

	col_sum		
15	8	11	14

	col_prod		
45	0	21	64

row_max
4
8
9

overall_min

0

row_indx_max

4

4

1

row_indx_min

1

1

2

col_1_3mean

5 3.6666667

col_ssq

107   40   59   84

mean_col_max

7.5

2.2.2 Element-wise Operators

Element-wise operators perform mathematical operations on individual elements of a matrix. The following is a list of IML element-wise operators.

Operator	Action
/	division operator
<>	element maximum operator
><	element minimum operator
#	element-wise multiplication operator
##	element-wise power operator

Example 2.8 – Element-wise operators:

The following code demonstrates uses of element-wise operators using the 5 × 1 column vectors **A** and **B**.

```
PROC IML;
    a={1,2,3,4,5};
    b={5,4,3,2,1};
    div_ab=a/b;
    max_ab=a<>b;
```

```
    min_ab=a><b;
    mult_ab=a#b;
    pow_ab=a##b;
    PRINT a b div_ab max_ab min_ab mult_ab pow_ab;
QUIT;
```

The statement

```
    div_ab=a/b;
```

divides each element in **A** with the corresponding element in **B**. The statement

```
    max_ab=a<>b;
```

compares each element in **A** with the corresponding element in **B** and returns the maximum. The statement

```
    min_ab=a><b;
```

compares each element in **A** with the corresponding element in **B** and returns the minimum. The statement

```
    mult_ab=a#b;
```

multiplies each element in **A** with the corresponding element in **B**. The statement

```
    pow_ab=a##b;
```

raises each element in **A** to the power of the corresponding element in **B**.

Output – Example 2.8:

a	b	div_ab	max_ab	min_ab	mean_ab	pow_ab
1	5	0.2	5	1	5	1
2	4	0.5	4	2	8	16
3	3	1	3	3	9	27
4	2	2	4	2	8	16
5	1	5	5	1	5	5

2.2.3 Comparison Operators

Comparison operators are used to compare two literals, matrices, or equations. Evaluations containing comparison operators involve binary responses: 1 if the comparison is true and 0 if the comparison is false. Comparison operators are often used with control statements for computational statements to occur or not depending on whether or not a comparison is true. The following is a list of IML comparison operators.

Operator	Action
<	less than
<=	less than or equal to
=	equal to
>	greater than
>=	greater than or equal to
^=	not equal to

It is important to note that the SAS DATA step allows the abbreviations GE, GT, LE, and LT to represent the operators >=, >, <=, and < respectively. The IML procedure does not allow the abbreviations and will generate an error if they are used. It is also important to note that compound inequalities are not valid in the IML procedure. For example, the code

```
IF low <= middle <= high THEN within="Y";
```

is valid in a SAS DATA step, but not in a PROC IML step. However, the PROC IML step would accept the following:

```
IF (low <= middle) & (middle <= high) THEN within="Y";
```

Example 2.9 – Comparison operators:

Ten average daily temperature measures are recorded and compared to the temperature of interest, 72. The following code creates column vectors using each of the comparison operators. The resulting element in each row of each column vector is a 1 if the comparison is true for that element and 0 if the comparison is false.

```
PROC IML;
   a={70,70,71,74,72,78,77,75,73,74};
   low=a[,1]<72;
   high=a[,1]>72;
   equal=a[,1]=72;
   not_equal=a[,1]^=72;
   ge=a[,1]>=72;
   le=a[,1]<=72;
   PRINT low high equal not_equal ge le;
QUIT;
```

Output – Example 2.9:

low	high	equal	not_equal	ge	le
1	0	0	0	0	1
1	0	0	0	0	1
1	0	0	0	0	1
0	1	0	0	1	0
0	0	1	1	1	1
0	1	0	0	1	0
0	1	0	0	1	0
0	1	0	0	1	0
0	1	0	0	1	0
0	1	0	0	1	0

2.2.4 Index Creation Operator

There is only one index creation operator. The index creation operator (:) creates a row vector with elements including all real numbers between two specified values in ascending order (if the first listed element is less than the second listed element) or descending order (if the first listed element is greater than the second listed element).

Example 2.10 – Index creation operators:

The following code demonstrates the index creation operator, creating an increasing series of numbers and a decreasing series of numbers.

```
PROC IML;
   a=1:5;
   b=5:1;
   PRINT a,b;
QUIT;
```

Output – Example 2.10:

		a		
1	2	3	4	5

		b		
5	4	3	2	1

2.2.5 Logical Operators

The following logical operators may also be used in conjunction with control statements:

Operator	Action
&	and
|	or

The SAS DATA step allows the operators AND and OR to be used in logical expressions. The PROC IML step does not. The symbolic representations & and | must be used for logical operations in the PROC IML step. The following example demonstrates the use of comparison operators and logical operators.

Example 2.11 – Comparison and logical operators:

This example checks to see if, for each row, the number in the first column lies within the numbers in the second and third columns of the matrix **M**.

```
PROC IML;
   m = {1 2 3,
        4 2 6,
        5 1 9,
        3 7 8};
   int = J(4,1,"N");
   int[LOC((m[,2] <= m[,1]) & (m[,1] <= m[,3]))] = "Y";
   PRINT m int;
QUIT;
```

First, the 4×3 matrix **M** is defined. Next, the 4×1 column vector **INT** is defined using the function J, creating a vector with the character string "N" populating each element. This initializes **INT**. Then, the statement

```
   int[LOC((m[,2] <= m[,1]) & (m[,1] <= m[,3]))] = "Y";
```

checks to see if the decreasing order of the elements in each row of the matrix **M** is $M_{i,2} \leq M_{i,1} \leq M_{i,3}$, for each of the rows, $i = 1$ to 4. If that is the case, the value of **INT** for that row changes to "Y". If not, it remains "N". It is interesting to note the use of the LOC function that in this example prevents unnecessary looping. By not providing a row reference in the brackets, **M[,1]**, **M[,2]**, and **M[,3]**, include all rows resulting in location referencing of all rows in the three column vectors. At each row, the combination of comparison and logical operators checks the condition of the element in the first column being contained within the second and third columns. Lastly, the PRINT statement prints **M** and **INT** to the output.

Output – Example 2.11:

M			INT
1	2	3	N
4	2	6	Y
5	1	9	Y
3	7	8	N

2.3 Chapter Exercises

2.1 Describe, compare, and contrast each of the following terms as they apply to SAS/IML:

    function, operator, call, command, control statement

2.2 Explain the difference between subscript reduction operators and element-wise operators as they apply to SAS/IML.

2.3 Explain the difference between comparison operators and logical operators.

2.4 A researcher is interested in printing out the status, "healthy" or "unhealthy" for each of nine subjects based on their BMI score. A healthy BMI score is one that is between 18.5 and 25. Consider the following code:

```
PROC IML;
    bmi={15,22,27,19,32,17,23,31,20};
    IF 18.5<= bmi <= 25 THEN status="healthy";
    ELSE status="unhealthy";
    PRINT status;
QUIT;
```

There are several coding issues that prevent the above code from printing the status for each of the nine subjects. Fix the code so that it works as intended. Use the LOC function in the code.

2.5 A researcher is interested in printing out the status for each of nine subjects based on their BMI score according to the following criteria: A BMI score below 17.5 represents a status of "severely underweight", a BMI score at or above 17.5 and below 18.5 represents a status of "underweight", a BMI score at or above 18.5 and at or below 25 represents a status of "healthy", a BMI score above 25 and at or below 30 represents a status of "overweight", and a BMI score above 30 represents a status of "severely overweight". Consider the following code:

```
PROC IML;
  bmi={15,22,27,19,32,17,23,31,20};
  IF 18.5<= bmi<= 25 THEN status="healthy";
  ELSE status="unhealthy";
  PRINT status;
QUIT;
```

There are several coding issues that prevent the above code from printing the status for each of the nine subjects. Fix the code so that it works as intended.

Chapter 3
IML Programming Features

3.1 Modules

A SAS/IML module is a subprogram or subroutine. A module is set apart from the rest of the IML code and is not executed until called. Modules can be called various times within a single PROC IML step. This allows the user to consolidate code by writing repeatedly-used code once as a module as opposed to repeating similar code various times, making the coding more clean and efficient. A module can then be called repeatedly as opposed to typing large sections of code repeatedly. Modules can also be saved using a STORE MODULE statement and then reloaded using a LOAD MODULE statement. A module may be nested within another module, but cannot recursively call itself.

3.1.1 Defining and Executing Modules

A module starts with a START statement and ends with a FINISH statement. A module is executed using either a RUN statement or a CALL statement. The difference between a RUN statement and a CALL statement are only manifest when a module shares its name with a built-in IML subroutine. If that is the case, the CALL statement should be used to execute the built-in IML subroutine and the RUN statement should be used to execute the module. It is also possible to pass arguments to a module and to have a module compute results and produce output. A module may operate just like a built-in IML function or CALL subroutine.

The following example demonstrates the creation and execution of a module.

Example 3.1 – Computing the rank of a matrix:

This example demonstrates the use of a module to create a function that will accept an input matrix and compute the rank of that matrix.

There are multiple ways of computing the rank of a matrix. One method is the following:

$$rank(A) = TR(A^-A)$$

This method requires the computation of the MP-inverse (see Section 5.1.5) of the matrix **A**. It is advantageous to find a method for computing the rank of a matrix that does not involve the computation of the MP-inverse. A second method for computing the rank of the matrix **A** involves reducing the matrix **A** to row echelon form and then counting the number of rows in the resulting matrix with non-zero elements. Consider the following code:

```
PROC IML;
    a={1 2 3,
       4 5 6,
       7 8 9};
    START mat_rank(mat);
        e=ECHELON(mat);
        e=(e^=0)[, +];
        mat_rank=(e^=0)[+, ];
        RETURN(mat_rank);
    FINISH;
    rank=mat_rank(a);
    PRINT a rank;
QUIT;
```

The function MAT_RANK is created using modular programming by the same name. The coding for the module starts with the statement

```
    START mat_rank(mat);
```

The module is named MAT_RANK and the matrix **MAT** is passed to the module. The statement

```
    e=ECHELON(mat);
```

creates the matrix E by reducing the input matrix **MAT** to row echelon form using the IML function ECHELON. The statement

```
    e=(e^=0)[, +];
```

creates a column matrix **E** by summing the number of non-zero elements across each row of the row echelon reduced matrix **E** (note that the current assignment to the matrix **E** overwrites its previous assignment). The statement

```
    mat_rank=(e^=0)[+, ];
```

creates a scalar MAT_RANK by summing the number of non-zero elements for each row of the column vector **E**. The result is the rank of the input matrix **MAT**. The statement

```
    RETURN(mat_rank);
```

indicates the module MAT_RANK is to be a function by the same name. The module returns the value of the scalar MAT_RANK as the result. The FINISH statement ends the modular code. The newly created function MAT_RANK computes the rank of a matrix and does not require matrix inversion as a step in the process to obtain the rank.

The statement

```
rank=mat_rank(a);
```

computes the rank of the matrix **A** using the function MAT_RANK and assigns the value to the variable **RANK**. The PRINT statement prints the matrix **A** and scalar **RANK** to the output.

Output – Example 3.1:

```
              a                              rank

     1         2         3                     2
     4         5         6
     7         8         9
```

3.1.2 Symbol Tables

Working with modules involves symbol tables. Within a PROC IML step, scalars, vectors, and matrices can be defined in assignment statements. Within the IML computing environment a module can be defined. Computing that occurs between the START and FINISH statements of a module are considered within the *module computing environment* or *module environment*. A global symbol table maintains the values of the variables (symbols) that have been created within the IML computing environment in *immediate mode* (outside a module) for the current PROC IML step. Assignment of values to variables within the module environment also affects the global symbol table as long as no arguments are passed to the module. If values are passed to the module, a local symbol table is created. All assignments made within that module are assignments to the local symbol table and each module will have its own local symbol table.

Arguments can be passed to a module. These arguments may include literals, matrices, subscripted matrices, and expressions. Arguments passed to a module may include *output variables*. An output variable is passed to a module and then assigned a value within the module. The output variable creates an entry on both the local symbol table for the module as well as the global symbol table. Consequently, the output variable is recognized outside the module whereas variables stored only on the local symbol table are only recognized within the module. Arguments passed to a module can include both output variables and local variables.

The values assigned within one symbol table need not be maintained within another symbol table. For example, a variable X may be assigned a value in the global table. Another variable with the same name, X, may be assigned in the local environment for a module. A third variable with the same name, X, may be assigned in the local environment for a second module. There need not be any connection between any of the three X variables in the three symbol tables.

Consider the following example:

Example 3.2 – Symbol tables:

Within the PROC IML step there are four assignments of the variable **X**. The first assignment occurs in immediate mode prior to the first module START statement. That assignment enters **X** with a value of 5 into the global symbol table for the PROC IML step. Three modules are created, MODULE1, MODULE2, and MODULE3. There are no arguments passed to MODULE1 and so the assignment of **X**=7 made within that module is an assignment in the global symbol table and over-writes the previous value of **X**. The second module, MODULE2, is passed the literal argument 3. That assigns the value 3 to the variable **X** in the local symbol table for MODULE2 and does not affect the value of **X**=7 in the global symbol table. The creation of the variable **Y** within MODULE2 is another assignment within the local symbol table for MODULE2. No value exists for the variable **Y** within the global symbol table. Consequently, the statement

```
PRINT "first occurrence of y" y;
```

creates an error message in the log indicating that **Y** has not been set to a value. The third module, MODULE3, is passed the literal argument 1 to the local variable **X** and defines an output variable **Y**. The value **X**=1 and assignment of **Y**=8 affect the local symbol table within MODULE3. This local symbol table is not the same as the local symbol table for MODULE2. Because **Y** is identified as an output variable, its value, **Y**=8, affects the global symbol table as well as the module's local symbol table. Consequently, the statement

```
PRINT "second occurrence of y" y;
```

produces output.

```
PROC IML;
   x=5; ** assignment within the global symbol table;
   START module1;
      x=7; ** assignment within the global symbol table;
   FINISH;
   START module2(x);
      y=2; ** assignment within the local symbol table for
         module2;
   PRINT x y;
   FINISH;
   START module3(y,x);
      y=8; ** assignment within the local symbol table for
         module2;
   FINISH;
   PRINT x;
   RUN module1;
   PRINT x;
```

```
    RUN module2 (3); ** local assignment of x=3;
    PRINT x;
    PRINT "first occurrence of y" y;
    RUN module3 (y,1); ** local assignment of x=1, output
        variable y;
    PRINT "second occurrence of y" y;
QUIT;
```

Output – Example 3.2:

```
                        x
                        5

                        x
                        7

              · x                y
                3                2

                        x
                        7
                                             y
              second occurrence of y         8
```

The first value of **X** printed, **X**=5, is a result of the first PRINT statement not contained within a module. Although the modules occur prior to that PRINT statement, the execution of the modules does not occur until they are executed through the RUN statement. The variable **X** obtains its value, 5, at the beginning of the PROC IML step.

The statement

```
    RUN module1;
```

executes the code for MODULE1. This module is not passed arguments and so modifies the global symbol table with the assignment statement

```
    x=7; ** assignment within the global symbol table;.
```

As a result of the overwriting of the value of **X** with the value 7, the second PRINT statement prints the value of **X**=7 to the output. The statement

```
    RUN module2 (3); ** local assignment of x=3;
```

executes the code for MODULE2. This module is passed the literal argument 3 which is assigned to the local variable **X**. Because arguments are passed to the module, a local symbol table is created and variable assignments within the module, the assignment of the value of 3 to the variable **X** and the assignment of the value 2 to the variable **Y**, impact the local symbol table and not the global symbol table. Consequently, the statement

```
PRINT x y;,
```

contained within the module, recognizes the local symbol table and prints the values **X**=3 and **Y**=2. The statements

```
PRINT x;
PRINT "first occurrence of y" y;
```

which occur outside the module refer to the global symbol table. The global symbol table still holds the value **X**=7 and does not recognize the variable Y that is defined within the local symbol table within MODULE2. The statement

```
PRINT "first occurrence of y" y;
```

produces no output, but sends an error message to the log indicating that **Y** has not been set to a value (on the global symbol table). The statement

```
RUN module3(y,1); ** local assignment of x=1,
ouput variable y;
```

executes the code for MODULE3. The variable **Y** is set up as an output variable. Consequently, its assignment, **Y**=8, within the module creates an entry in the local symbol table as well as the global symbol table. The literal argument 1 is passed to the variable **X** in the local symbol table for MODULE3. The global symbol table still holds the value 7 for the variable **X**. The statement

```
PRINT "second occurrence of y" y;
```

prints the value **Y**=8 to the output.

3.2 Output Options

The IML procedure can generate a variety of matrices, data sets, text files, and output. There are different ways of dealing with these different forms of output. Three will be discussed here: creating SAS data sets from matrices, sending output to an output file (or output window in the windows environment), and sending output to a text, RTF, HTML, or PDF file.

3.2.1 Creating a SAS Data Set from a Matrix

The two most common headaches encountered by the author with IML is remembering how to convert SAS data sets into matrices, and how to convert matrices into SAS data sets. The prior is explained in section 1.9.2, and the latter is as follows:

Consider the following 5×3 matrix, $\mathbf{A}$:

$$\mathbf{A} = \begin{bmatrix} 10 & 88 & 92 \\ 9 & 91 & 95 \\ 10 & 85 & 83 \\ 8 & 90 & 87 \\ 10 & 92 & 96 \end{bmatrix}$$

The following code can be used to convert this matrix into a data set with three variables and five observations.

Example 3.3 – Creating a data set from a matrix:

The following code creates the data set DSET_A from the matrix $\mathbf{A}$:

```
PROC IML;
   a={10 88 92,9 91 95,10 85 83,8 90 87,10 92 96};
   CREATE dset_a FROM a [COLNAME={a1,a2,a3}];
   APPEND FROM a;
QUIT;

PROC PRINT DATA=dset_a;
RUN;
```

Output – Example 3.3:

Obs	A1	A2	A3
1	10	88	92
2	9	91	95
3	10	85	83
4	8	90	87
5	10	92	96

The

```
[COLNAME={a1,a2,a3}]
```

option gives the variable names **A1**, **A2**, and **A3** for the three columns of the newly created data set DSET_A. If unspecified, the column names will take the generic form **COL1**, **COL2**, and **COL3**.

It is also possible to create a data set that includes multiple variables (column vectors) from more than one source vector. For example, the following example creates a data set from two column vectors.

Example 3.4 – Creating a data set from two column vectors:

The following code creates the data set CLASSDAT from the column vectors **NAMES** and **SCORES**:

```
PROC IML;
   names={"Rob","Joe","Erich","Lisa","Mike"};
   scores={92,95,83,87,96};
   CREATE classdat VAR{Names Scores};
   APPEND;
QUIT;

PROC PRINT DATA=classdat;
RUN;
```

Output – Example 3.4:

Obs	NAMES	SCORES
1	Rob	92
2	Joe	95
3	Erich	83
4	Lisa	87
5	Mike	96

The variable **SCORES** in this data set is numeric as a result of being created from a numeric column vector. Had the data set been created from a character matrix, all variables would be characters – even if some of the columns contained only numbers. It is important to note that this method only works for column vectors. One could create a data set from three or more column vectors using the same general method, and all variables will retain the original vector type (character or numeric).

3.2.2 Sending Data to the Output

Resulting matrices and other information generated by a PROC IML step can be sent to the output to be viewed. When using batch mode in the UNIX environment, these results are sent to a text file with the name being that of the program, and the extension being .LST. In the WINDOWS environment or the window-type interface to the UNIX version of SAS, the results are sent to a window designated as the OUTPUT window.

The most common way to send data to the output is with the PRINT statement. The PRINT statement sends the contents of a matrix, vector, or scalar to the output. The standard format of a PRINT statement is

PRINT *<matrices><(expression)>* <"*message*"> *<pointer-controls>* *<[options]>*;.

The input, *matrices,* is the name of one or more matrices, vectors, or scalars defined within the current IML procedure. An *expression* contained within parentheses is evaluated and printed. The *messages* input allows for text to be printed. The pointer-controls include commas (,) and slashes (/). Output data that are separated by a space are printed on the same line (space permitting). Output data that are separated by a comma (,) are printed on separate lines. Output data that are separated by a slash (/) are printed on separate pages. The *options* include the following:

- COLNAME=*matrix* uses the values of a character matrix to label the columns of the matrix to be printed.
- ROWNAME=*matrix* uses the values of a character matrix to label the rows of the matrix to be printed.
- FORMAT=*format* specifies the format to be applied to the elements of the matrix to be printed.

The following example demonstrates the use of the PRINT statement.

Example 3.5 – The PRINT command:

This example contains the number of students in an introductory course of statistics grouped by gender.

```
PROC IML;
   gender={Males,Females};
   totals={21 0.4666666667,24 0.5333333333};
   PRINT gender totals (totals[1,1]);
QUIT;
```

Output – Example 3.5:

```
                  GENDER      TOTALS

                  MALES           21 0.4666667
                  FEMALES         24 0.5333333
```

The statement

```
PRINT gender totals(totals[1,1]);
```

prints the column vector **GENDER**, the matrix **TOTALS**, and the element of the first row and first column of the matrix **TOTALS**. Note that expressions included within parentheses do *not* have to be defined matrices, but simply valid expressions. When the PRINT command is used to print a matrix, by default, the name of the matrix appears above the matrix in the output. However, no label appears, by default, above an expression.

   If the output of is not comprehensive enough, there are many tools that allow it to be modified. This next example has the same numeric results, but it uses various tools to change the appearance of the output.

Example 3.6 – Formatting output:

This example contains the number of students in an introductory course of statistics grouped by gender.

```
OPTIONS NODATE NONUMBER;
PROC IML;
   TITLE "Intro. Stats. Demographics";
   TITLE2 "Males vs. Females";
   totals={21 0.4666666667,24 0.5333333333};
   rows={"Males","Females"};
   cols={"     Totals","Percentages"};
   RESET FW=4 SPACES=2;
   PRINT(totals[1:2,1]) [COLNAME=(cols[1,1]) ROWNAME=rows]
        (totals[1:2,2]) [COLNAME=(cols[2,1]) FORMAT=PERCENT.];
QUIT;
```

Output – Example 3.6:

```
                    Intro. Stats. Demographics

                        Males vs. Females

                          Totals   Percentages
            Males            21         47%
            Females          24         53%
```

Many formatting tools went into this example. The

```
OPTIONS NODATE NONUMBER;
```

statement suppresses the date and page number from appearing in the output. The two TITLE statements provide title and subtitle.

The

```
RESET FW=4 SPACES=2;
```

statement has IML print the matrices, **TOTALS**, with two pad spaces between each element (the default is nine pad spaces) and four spaces between each of the matrices that appear on the same row (the default is one space).

The PRINT statement allows the printing of expressions as well as defined matrices. To print an expression, surround it with parentheses, as with

```
(totals[1:2,1]).
```

This allows the printing of specified rows and columns as opposed to printing an entire matrix. The COLNAME and ROWNAME options provide detailed row and column labels. The FORMAT command allows for representation of the decimal values in percentage form. There are many SAS formats that can be used with the FORMAT command. These SAS formats can be grouped in three categories: character, date and time, and numeric. An example of basic SAS formats that can be used with the FORMAT= option in the PRINT statement in a PROC IML step is represented in the following table:

Table 3.1 Various SAS formats that can be used with the FORMAT= option in the PRINT statement in a PROC IML step, their names, and the result they have when applied to values

Input Value	Format	Result
hello world	$UPCASE11.	HELLO WORLD
123456789	SSN.	123-45-6789
12345	MMDDYY10.	10/19/1993
12345	WORDDATE18.	October 19, 1993
12345	TIME7.	3:25:45
123456.789	DOLLAR11.	$123,456.79
123456.789	10.3	123456.789
123456.789	COMMA10.3	123,456.789

3.2.3 Sending Data to a Text File

The easiest way to send the results of a PROC IML step or any other PROC step to a text file is to use the PRINTTO procedure. In essence, it is a two-step process.

Step 1: Include the following PROC PRINTTO step, including the location to where the output is to be sent. This PROC step should precede the PROC IML step that generates the output.

```
PROC PRINTTO PRINT='location';
RUN;
```

Step 2: Include the following PROC PRINTTO step. This PROC step should follow the PROC IML step that generates the output. It restores the output settings to their default.

```
PROC PRINTTO;
RUN;
```

The following example demonstrates how the PRINTTO procedure is used to send output from the IML procedure to a text file.

Example 3.7 – Sending output to a text file:

This example generates the same output as did the previous example. However, in this example the output is sent to the text file C:\IMLFILE.TXT instead of to the default output location.

```
PROC PRINTTO PRINT='c:\imlfile.txt';
RUN;

OPTIONS NODATE NONUMBER;
PROC IML;
   TITLE "Intro. Stats. Demographics";
   TITLE2 "Males vs. Females";
   totals={21 0.4666666667,24 0.5333333333};
```

```
rows={"Males","Females"};
cols={"       Totals","Percentages"};
RESET FW=8 SPACES=2;
PRINT(totals[1:2,1]) [COLNAME=(cols[1,1]) ROWNAME=rows]
      (totals[1:2,2]) [COLNAME=(cols[2,1]) FORMAT=PERCENT.];
QUIT;

PROC PRINTTO;
RUN;
```

The following output is sent to the text file C:\IMLFILE.TXT.

Output (sent to the text file C:\IMLFILE.TXT) – Example 3.7:

```
                Intro. Stats. Demographics
                     Males vs. Females

                          Totals  Percentages
                 Males       21         47%
                 Females     24         53%
```

3.2.4 Sending Data to RTF, HTML, and PDF Files via ODS

Starting with version 7 and continuing with versions 8 and 9, the SAS System has made use of the ODS (Output Delivery System), which allows more flexibility to program output. The ODS sends output along with formatting instructions to a variety of possible destinations including RTF (Rich-Text Format), HTML (Hyper Text Markup Language), and PDF (Portable Document Format) files. A benefit of the RTF output option is that resulting output is easily incorporated into Microsoft Word documents. Similarly, the HTML output option allows for resulting output to be easily incorporated into web pages for view on intra- and internets. The PDF format is widely used for document presentation on the internet.

The following example demonstrates how the ODS is used to send output from the IML procedure to RTF, HTML, and PDF files.

Example 3.8 – Sending output to RTF, HTML, and PDF files:

The code for this example generates output in different file formats sent to different file locations. The output is sent to the RTF file C:\IMLFILE.RTF, the HTML file C:\IMLFILE.HTML, and the PDF file C:\IMLFILE.PDF instead of to the default output location.

```
OPTIONS NODATE NONUMBER;

ODS RTF FILE='c:\imlfile.rtf';
ODS HTML FILE='c:\imlfile.html';
ODS PDF FILE='c:\imlfile.pdf';
```

```
PROC IML;
   TITLE "Intro. Stats. Demographics";
   TITLE2 "Males vs. Females";
   totals={21 0.4666666667,24 0.5333333333};
   rows={"Males","Females"};
   cols={"      Totals","Percentages"};
   RESET FW=8 SPACES=2;
   PRINT (totals[1:2,1]) [COLNAME=(cols[1,1]) ROWNAME=rows]
         (totals[1:2,2]) [COLNAME=(cols[2,1]) FORMAT=PERCENT.];
QUIT;

ODS PDF CLOSE;
ODS HTML CLOSE;
ODS RTF CLOSE;
```

The

```
ODS RTF FILE='c:\imlfile.rtf';
```

statement opens the RTF destination and assigns the output to the destination file C:\IMLFILE.RTF. The

```
ODS HTML FILE='c:\imlfile.html';
```

statement opens the HTML destination and assigns the output to the destination file C:\IMLFILE.HTML. The

```
ODS PDF FILE='c:\imlfile.pdf';
```

statement opens the PDF destination and assigns the output to the destination file C:\IMLFILE.PDF. The

```
ODS RTF CLOSE;,
ODS HTML CLOSE;,
```

and

```
ODS PDF CLOSE;
```

statements close the RTF, HTML, and PDF destinations. All output generated by the code between the

```
ODS destination FILE=file;
```

and

```
ODS destination CLOSE;
```

statements is sent to the destination file. In this example, output seen in Example 1.27 will be sent to the output window. Additionally, the RTF file C:\IMLFILE.RTF, the HTML file C:\IMLFILE.HTML, and the PDF file C:\IMLFILE.PDF are created.

Output – Example 3.8: Appearance of RTF and PDF Files

Intro. Stats. Demographics

Males vs. Females

	Totals	Percentages
Males	21	47%
Females	24	53%

The first output shows the appearance of the RTF file C:\IMLFILE.RTF and the PDF file C:\IMLFILE.PDF as they appear in an RTF browser such as Microsoft Word and a PDF browser such as Adobe Acrobat Reader, respectively.

Output – Example 3.8 (continued): Appearance of HTML File

Intro. Stats. Demographics
Males vs. Females

	Totals	Percentages
Males	21	47%
Females	24	53%

The second output shows the appearance of the HTML file C:\IMLFILE.HTML, as it appears in an HTML browser.

It is possible to use and modify templates to alter the appearance of the output.

Example 3.9 – Journal style:

The following example code is similar to that of the previous example. This example uses the STYLE=JOURNAL option to modify the appearance of the output to an appearance more consistent with that of published journal manuscripts.

```
OPTIONS NODATE NONUMBER;
ODS RTF FILE='c:\imlfile.rtf' STYLE=JOURNAL;
PROC IML;
    TITLE "Intro. Stats. Demographics";
```

```
TITLE2 "Males vs. Females";
totals={21 0.4666666667,24 0.5333333333};
rows={"Males","Females"};
cols={"     Totals","Percentages"};
RESET FW=8 SPACES=2;
PRINT (totals[1:2,1]) [COLNAME=(cols[1,1]) ROWNAME=rows]
      (totals[1:2,2]) [COLNAME=(cols[2,1]) FORMAT=PERCENT.];
QUIT;

ODS RTF CLOSE;
```

Output – Example 3.9:

Intro. Stats. Demographics
Males vs. Females

	Totals	Percentages
Males	21	47%
Females	24	53%

The STYLE=JOURNAL option can be applied to RTF, HTML, and PDF output destinations.

3.3 Documenting and Formatting Code

An important part of coding is documentation and code formatting, or "programming etiquette" as it may be called. The ability of an individual to take written code, decipher its meaning, debug errors, and modify the code for future alternate use is often highly dependent upon the code documentation and format.

3.3.1 Documenting Code

SAS has two specific features for documenting code: single-line documentation and multi-line documentation. For single line documentation, a statement begins with an asterisk (*) and ends with a semicolon. Everything contained within such a statement is not considered executable code and is essentially ignored by the compiler, thus serving as notation to the program user.

Multi-line documentation occurs when the statements are offset with /* and */ (forward slash and asterisk). Everything within those markers is ignored by the compiler. Multi-line documentation is used when multiple lines of documentation appear together. It does not require the user to include an asterisk at the beginning and a semicolon at the end of each documentation statement.

Example 3.10 – Single- and multi-line documentation:

The following code appears in Example 1.10. It appears again here to demonstrate the concepts of single- and multi-line documentation. Because the purpose of this example is to demonstrate the documentation, the output is not provided, but can be had in Example 1.10.

```
/**************************************************/
/** SINGLE- AND MULTI-LINE DOCUMENTATION EXAMPLE **/
/**************************************************/
/** The following code demonstrates uses of the **/
/** single- and multi-line documentation.       **/
DATA class_b;
   FILENAME inclass 'a:\class data\class b.txt';
   INFILE inclass MISSOVER;
   INPUT student $ score;
RUN;
/*
PROC PRINT DATA=class_b;
RUN;
*/
PROC IML;
   USE class_b; * opens the data set CLASS_B for use;
   READ ALL VAR {student} INTO students;
   READ ALL VAR {score} INTO scores;
   CLOSE class_b; * closes the data set CLASS_B;
   PRINT students scores; ** prints newly-created column
      vectors **;
QUIT;
```

Prior to executable code, the section

```
/**************************************************/
/** SINGLE- AND MULTI-LINE DOCUMENTATION EXAMPLE **/
/**************************************************/
```

gives a title to the program that is easily seen as it is surrounded by many asterisks. This provides an example of one use of the multi-line documentation—offsetting a title. It is good practice to include a title or description of a program for future use. A researcher whose programs look very similar may have trouble distinguishing programs simply by the file name when wanting to find previously-written code for a future application.

The statements

```
/** The following code demonstrates uses of the **/
/** single- and multi-line documentation.       **/
```

provide another example of how multi-line documentation may be used.

The statements

```
/*
PROC PRINT DATA=class_b;
RUN;
*/
```

demonstrate a method for skipping the execution of a section of code within a program. In addition to creating comments, multi-line documentation methods can be used to prevent code from compiling without deleting the code. All code within the markers /* and */ are skipped by the compiler when the program is executed. A use for this is debugging code.

The statement

```
* opens the data set CLASS_B for use;
```

demonstrates a basic use of single-line documentation. It follows the

```
USE class_b;
```

statement on the same line. Because the single-line documentation begins with an asterisk and ends with a semi-colon, it is not considered code by the compiler. The documentation could have also appeared on a separate line of the code. The statement

```
** prints newly-created column vectors **;
```

demonstrates that a combination of asterisks can be creatively used make the documentation stand out more in the code.

3.3.2 Use of Case and Spacing

Programs can include dozens of lines of code. Working with dozens of lines of code can be difficult. It can be difficult to identify the beginnings and endings of PROC and DATA steps, nested loop boundaries, etc. It can be difficult to distinguish variable names from functions. The use of case and spacing can help organize code so that it is easier to read and manage. Vertical spacing can be used to separate PROC and DATA steps. Horizontal spacing can be used to identify beginnings and endings of loops. Using all capital-letters for built-in SAS functions and designations and all lower-case letters for variable names can help distinguish the two.

Example 3.11 – Use of case and spacing:

The following two examples show the code from the previous example. The second set of code shows the use of case and spacing and the first set of code does not. Because the purpose of this example is to demonstrate the use of case and spacing, the output is not provided, but can be had in Example 1.10.

Code without use of case and spacing

```
data class_b;
filename inclass 'a:\class data\class b.txt';
infile inclass missover;
input student $ score;
run;
proc print data=class_b;
run;
proc iml;
use class_b;
read all var {student} into students;
read all var {score} into scores;
close class_b;
print students scores;
quit;
```

Code with the use of case and spacing

```
DATA class_b;
   FILENAME inclass 'a:\class data\class b.txt';
   INFILE inclass MISSOVER;
   INPUT student $ score;
RUN;

PROC PRINT DATA=class_b;
RUN;

PROC IML;
   USE class_b;
   READ ALL VAR {student} INTO students;
   READ ALL VAR {score} INTO scores;
   CLOSE class_b;
   PRINT students scores;
QUIT;
```

The second section of code is less difficult to read and manage than the first section of code. In the second section of code, vertical spacing separates the one DATA step and two PROC steps. Horizontal spacing clearly shows clearly the beginning and ending of the DATA step and PROC steps. Use of case clearly distinguishes built-in SAS designations from those determined by the user.

There are clearly multiple methods for documenting and organizing code. There is not one right way. However, it is important that the user take into consideration the possibility that code currently being written may need to be reviewed in the future. The more well organized and documented the code, the easier it will be to review the code in the future.

3.4 Chapter Exercises

3.1 Create a module named DIMENSIONS that is passed an input object (scalar, vector, or matrix) and will output the name and dimensions of the object. Demonstrate its use on a scalar, vector, and matrix.

3.2 Create a modular function named

```
STANDARDIZE (mat,mean,stdev)
```

that is passed three input objects: **MAT**, an input matrix of any dimensions; **MEAN**, a scalar constant as a mean; and **STDEV**, a scalar constant for the standard deviation. The function will standardize all the values in the input matrix, **MAT**, by subtracting **MEAN** from each element and then dividing the resulting difference by **STDEV**. Test the function by passing input matrices of different dimensions and differing values for **MEAN** and **STDEV**. Test it on the following inputs: mat={1 2 3, 2 4 6}; mean=2; stdev=4;

3.3 The following code demonstrates the basic principle of the identity matrix:

```
PROC IML; A={1 2 3,4 5 6,7 8 9}; I=I (NROW (A)); IA=I*A;
   AI=A*I; PRINT A I IA AI; QUIT;
```

Modify the code so that it includes proper spacing and indentation. Include a detailed description of the program (at the beginning of the code) using multi-line documentation and label each object assignment using single-line documentation.

3.4 Write a program that uses five different IML functions do describe the matrix **A**, where the matrix A is defined as

$$A = \begin{bmatrix} 1\,2\,3 \\ 2\,3\,4 \\ 3\,4\,5 \end{bmatrix}.$$

At least one of the five functions should be user-defined through a module. Redirect the output to an external file with PDF format. The output should include a detailed title and appropriate labels for each output value.

3.5 Create a module named "row" that checks to see if a given vector is a row or column vector, and returns a row vector in either case. Test the module using the vectors

$$C = \begin{bmatrix} 1 \\ 2 \\ 3 \end{bmatrix},$$

and

$$D = \begin{bmatrix} 4\,5\,6 \end{bmatrix}$$

to see if in both cases, a row vector results.

3.6 Your own module:

    a. Create a module that you feel might be useful to researchers in your field of study.

    b. Include a description of what your module does and what it can be used for.

    c. Include comments that describe each step in your module.

    d. Demonstrate the use of your module using data obtained from a published source (say, example data from a textbook). Reference the source from which the data were obtained.

Chapter 4
Matrix Manipulations in SAS/IML

4.1 Accessing Different Sections of a Matrix

IML allows different levels of matrix access: element access, row access, column access, submatrix access.

4.1.1 Accessing a Single Element of a Matrix

The following examples demonstrate how to access and replace a single element of a matrix:

Example 4.1 – Accessing a single element of a matrix:

The following code assigns the variable **CELL23** the contents of the element in the (2,3) position of matrix **A**:

```
PROC IML;
   a={10 88 92,9 91 95,10 85 83,8 90 87,10 92 96};
   cell23 = a[2,3];
   PRINT cell23;
QUIT;
```

Output – Example 4.1:

cell23

95

Example 4.2 – Replacing a single element of a matrix:

The next example puts the contents of the variable **CELL23**, the number 100, into the (2,3) element of the matrix **A**:

```
PROC IML;
   a={10 88 92,9 91 95,10 85 83,8 90 87,10 92 96};
   PRINT "Prior to replace:" a;
   cell23 = 100;
```

```
    a[2,3] = cell23;
    PRINT "After replace:" a;
QUIT;
```

This also could have been accomplished directly by replacing the fourth and fifth lines of the previous code with

```
a[2,3] = 100;.
```

Output – Example 4.2:

	a		
Prior to replace:	10	88	92
	9	91	95
	10	85	83
	8	90	87
	10	92	96
	a		
After replace:	10	88	92
	9	91	100
	10	85	83
	8	90	87
	10	92	96

Notice the number 95 in the first matrix **A** that is replaced with the number 100 in the second matrix **A**.

4.1.2 Accessing a Matrix Row or Column

The following examples demonstrate how to access a row or a column of a matrix.

Example 4.3 – Accessing a row of a matrix:

The following code creates the 1×3 row vector **A_ROW2** from the 2nd row of the 5×3 matrix **A**:

```
PROC IML;
    a={10 88 92,9 91 95,10 85 83,8 90 87,10 92 96};
    a_row2 = a[2,];
    PRINT a a_row2;
QUIT;
```

Notice that by omitting the column index number from the reference A[2,] the reference applies to *all columns* of the matrix **A**.

Output – Example 4.3:

a			a_row2		
10	88	92	9	91	95
9	91	95			
10	85	83			
8	90	87			
10	92	96			

Example 4.4 – Accessing a column of a matrix:

The following code creates the 1×3 column vector **A_COL1** from the 1st column of the 5×3 matrix **A**:

```
PROC IML;
    a={10 88 92,9 91 95,10 85 83,8 90 87,10 92 96};
    a_col1 = a[,1];
    PRINT a a_col1;
QUIT;
```

Notice that by omitting the row index number from the reference A[,1] the reference applies to *all rows* of the matrix **A**.

Output – Example 4.4:

a			a_col1
10	88	92	10
9	91	95	9
10	85	83	10
8	90	87	8
10	92	96	10

4.1.3 Accessing a Submatrix

It may be of interest to identify a subsection of a matrix, a submatrix. The subsection may consist of a number of columns, consecutive or otherwise, from the original matrix; a number of rows, consecutive or otherwise, from the original matrix; or portions of a number of rows, indexed by a number of columns.

The square brackets ([and]) following the name of a matrix allow the user to access different rows/columns of that matrix. The format is

Matrix-name[<*starting-row : ending-row*><,><*starting-column : ending-column*>]

The row indices and column indices are separated by a comma (,). Numbers on the left of the comma represent row indices and numbers on the right of the comma represent column indices. Within a row or column index, a colon (:) separates the starting reference of the index from the ending reference of the index. For example,

A[1:3,3:4]

represents the first through the third row of the third and fourth column of the matrix A. When no colons are used, a single number can index a single row or column. The reference

A[1,2]

represents the element in the first row and second column of the matrix A. Multiple inconsecutive rows or columns can be referenced by using a space to separate the references in an index rather than the colon. Additionally, each index is contained within curly brackets ({ and }). The reference

A[{1 3},{1 4}]

represents the first and third rows of the first and fourth columns of the matrix A. Omitting an index references all rows or all columns. The reference

A[,1:2]

omits the row references and thus will reference all rows of columns 1 and 2 of the matrix A. Omitting all index references the entire matrix. The reference

A[]

is the same as the reference

A.

Consider the following example.

Example 4.5 – Accessing a submatrix of a matrix:

The following code demonstrates the creation of different submatrices from the matrix **A**.

```
PROC IML;
    a={10 88 92 6,9 91 95 4,10 85 83 8,8 90 87 10,10 92 96 1};
    a1334=a[1:3,3:4];
    a12=a[1,2];
    a1314=a[{1 3},{1 4}];
    ac12=a[,1:2];
    a_=a[];
    PRINT a,a1334 a12,a1314 ac12 a_;
QUIT;
```

Output – Example 4.5:

a

10	88	92	6
9	91	95	4
10	85	83	8
8	90	87	10
10	92	96	1

a1334

92	6
95	4
83	8

a12

88

a1314

10	6
10	8

ac12

10	88
9	91
10	85
8	90
10	92

a_

10	88	92	6
9	91	95	4
10	85	83	8
8	90	87	10
10	92	96	1

4.1.4 Working with Diagonal Matrices and Matrix Diagonals

The following example demonstrates the creation of a diagonal matrix and a matrix diagonal.

Example 4.6 – Accessing a diagonal matrix and a matrix diagonal:

The following code creates the 3×3 diagonal matrix **DIAG_B** and a 3×1 column vector, **VEC_B**, of the diagonal elements from the 3×3 matrix **B**, which is created from the 5×3 matrix **A**:

```
PROC IML;
   a={10 88 92,9 91 95,10 85 83,8 90 87,10 92 96};
   b = a[2:4,1:3];
   diag_b = DIAG(b);
   vec_b = VECDIAG(b);
   PRINT a b diag_b vec_b;
QUIT;
```

Output – Example 4.6:

a			b		
10	88	92	9	91	95
9	91	95	10	85	83
10	85	83	8	90	87
8	90	87			
10	92	96			

diag_b			vec_b	
9	0	0	9	
0	85	0	85	
0	0	87	87	

4.1.5 Elements of a Column Vector

The following example demonstrates how to access an element of a column vector.

Example 4.7 – Accessing an element of a column vector:

The following code creates the 5×1 column vector **K** and a scalar **K3** from the third element of the column vector **K**:

```
PROC IML;
    k={2,7,11,6,4};
    k3 = k[3];
    PRINT k k3;
QUIT;
```

Notice that when working with a column vector, it would be redundant to refer to an element by both its row index number and its column index number. There is only one column. Therefore, the row index number is all that is needed. The same *does not* apply when working with row vectors.

Output – Example 4.7:

K	K3
2	11
7	
11	
6	
4	

4.2 Performing Basic Mathematical Operations on Subsections of a Matrix

IML provides several operators that can be used to perform basic mathematical operations on elements of subsections of matrices. These operators are known as *subscript reduction operators* and are described in Chapter 2. These subscript reduction or mathematical operators include the following:

Operator	Action
+	addition
#	multiplication
<>	maximum
><	minimum
<:>	index of minimum
>:<	index of maximum
:	mean
##	sum of squares

The following example demonstrates how these mathematical operators can be used on subsections of a matrix:

Example 4.8 – Application of mathematical operators to submatrices:

This example performs mathematical operations on submatrices of the matrix **A**.

```
PROC IML;
   a={10 88 92,9 91 95,10 85 83,8 90 87,10 92 96};
   sum_c12 = a[+,1:2];
   prod_r23 = a[2:3,#];
   max_1213 = a[1:2,1:3][<>];
   PRINT a sum_c12 prod_r23, max_1213;
QUIT;
```

The scalar **SUM_C12** contians the sum of the elements of the first two columns of the matrix **A**. The scalar **PROD_R23** contains the product of the elements of the second and third rows of the matrix **A**. Consider the statement

```
   max_1213 = a[1:2,1:3][<>];
```

The a[1:2,1:3] portion of the statement defines a submatrix based on the first and second rows of the first and third columns of the matrix **A**. Because [<>] does not contain any row or column indices, this portion of the statement identifies the maximum of all elements of the submatrix identified by the a[1:2,1:3] portion of the statement. The 2×2 matrix **MAX_1213** contains the maximum of the elements for the first and second rows of the first and third columns of the matrix **A**.

Output – Example 4.8:

		a	
10	88	92	
9	91	95	
10	85	83	
8	90	87	
10	92	96	

sum_c12		prod_r23	max_1213
47	446	77805	95
		70550	

4.3 Special Matrices

IML is set up to simplify the creation of various matrices with special characteristics. For example, suppose 20×20 matrix of ones is needed. It would be very time consuming to type in all 400 ones by manual entry. Or suppose a large block-diagonal matrix for a mixed model application is needed. Fortunately, IML contains several functions designed to create these special matrices automatically.

The following examples, include some of the special matrices IML is set up to create. Additional IML functions for creating special matrices not included in this section include TOEPLITZ, HANKEL, HDIR.

Example 4.9 – Identity matrix, zero matrix, matrix of ones, and vector of ones:

The following code creates several patterned matrices and vectors:

```
PROC IML;
    n=5;
    m=3;
    mat = {1 2 3,4 5 6,7 8 9};
    im = I(m);
    jm = J(m);
    jnm = J(n,m);
    zero = J(m,n,0);
    jm_vec = J(m,1);
    PRINT im jm jnm zero jm_vec;
QUIT;
```

Output – Example 4.9:

IM			JM			JNM			ZERO					JM_VEC
1	0	0	1	1	1	1	1	1	0	0	0	0	0	1
0	1	0	1	1	1	1	1	1	0	0	0	0	0	1
0	0	1	1	1	1	1	1	1	0	0	0	0	0	1
						1	1	1						
						1	1	1						

These matrices are created using the I and J functions. The matrix **IM** is an identity matrix of dimension m. It is a square matrix with ones for the diagonal elements and zeros for the off-diagonal elements. The argument passed to the I function is a numeric value that determines the number of rows and columns in the resulting matrix. The matrix **JM** is a matrix of ones. If there is only one argument passed to the J function, the matrix will be a square matrix with the number of rows and columns equal to the argument. Such is the case with the matrix **JM**. However, if two arguments are passed to the function, the first argument represents the number of rows and the second argument represents the number of columns. Such is the case with the matrix **JNM**. If three arguments are passed to the J function, the third argument specifies the numeric or non-numeric value for each matrix element. Such is the case with the **ZERO** matrix. The 3×1 column vector **JM_VEC** is a column vector of ones. Each of these special matrices has uses in linear models.

Example 4.10 – Direct (Kronecker) product and block-diagonal matrix:

The direct product, also known as the Kronecker product, is a method of multiplication and expansion of a matrix. In general, **A@B**, which in some texts appears as A⊗B, multiplies every element of **A** by the entire matrix **B**. Using direct products can simplify linear models equations.

Common design matrices, for ANOVA models for example, are often set up as block-diagonal matrices. A block-diagonal matrix is a matrix made up of identical submatrices along the diagonal. The BLOCK function produces block-diagonal matrices.

Linear models matrices can be constructed using direct products and/or the BLOCK function. The following code simplifies the creation of matrices using direct products and block diagonals:

```
PROC IML;
   n=5;
   m=3;
   im = I(m);
   jm = J(m);
   jn = J(n);
   k1 = im@jm;
   k2 = BLOCK(jm,jn);
   PRINT k1,k2;
QUIT;
```

Output – Example 4.10:

k1

1	1	1	0	0	0	0	0	0
1	1	1	0	0	0	0	0	0
1	1	1	0	0	0	0	0	0
0	0	0	1	1	1	0	0	0
0	0	0	1	1	1	0	0	0
0	0	0	1	1	1	0	0	0
0	0	0	0	0	0	1	1	1
0	0	0	0	0	0	1	1	1
0	0	0	0	0	0	1	1	1

k2

1	1	1	0	0	0	0	0
1	1	1	0	0	0	0	0
1	1	1	0	0	0	0	0
0	0	0	1	1	1	1	1
0	0	0	1	1	1	1	1
0	0	0	1	1	1	1	1
0	0	0	1	1	1	1	1
0	0	0	1	1	1	1	1

The matrix **K1** is created by multiplying every element of **IM** by the matrix **JM**. The second matrix, **K2**, is a block-diagonal matrix. It is created using the BLOCK function and consists of the matrices, **JM**, and **JN** on the diagonals, with zeros filling in the off diagonals. Notice that using direct products produces a block-diagonal matrix with equal blocks. The BLOCK function allows for block-diagonal matrices with blocks of unequal sizes. Unequal-size blocks are necessary for unbalanced linear model designs.

Example 4.11 – Design matrices:

The DESIGN and DESIGNF functions can also be used to create design matrices for linear models applications. Each different number passed to the DESIGN function creates a new column in the resulting design matrix. Each number creates a row with a value in the column corresponding to the number's value. The DESIGNF function is similar to the DESIGN function but it can be used to produce full-rank design matrices. The following code demonstrates how certain design matrices can be created in IML:

```
PROC IML;
    x1 = DESIGN({1,1,1,2,2,2,3,3,3});
    x2 = DESIGNF({1,1,2,2,3});
    PRINT x1 x2;
QUIT;
```

Output – Example 4.11:

X1			X2	
1	0	0	1	0
1	0	0	1	0
1	0	0	0	1
0	1	0	0	1
0	1	0	-1	-1
0	1	0		
0	0	1		
0	0	1		
0	0	1		

Design matrices are used in ANOVA models. In some linear models and regression texts, the format of the **X1** matrix is referred to as "dummy coding" and the format of the **X2** matrix is referred to as "effect coding." The contents of the **X1** matrix might represent a design factor with three levels (three columns) and three measures at each level (three rows per level). The contents of the **X2** matrix might represent a design factor with three levels (one less than the number of columns) and two measures for each of the first two levels and one measure for the third level.

Example 4.12 – Hermite and echelon reduction matrices:

In linear models, the Hermite form of a matrix can be used in determining the *generalized inverse*. If

$$AA^gA = A \tag{4.1}$$

then A^g is a generalized inverse of A. Additionally, if 4.1 is true then

$$A^gA = H,$$

where **H** is the Hermite form of the matrix **A**.

Matrices may be reduced to row-echelon normal form in order to determine the rank of a matrix. The following code shows how to compute the Hermite form and the echelon-reduced form of a matrix:

```
PROC IML;
    RESET FUZZ;
    a={1 2 1,
       2 3 2,
       1 2 1};
    e=ECHELON(a);
    h=HERMITE(a);
    b=h*GINV(a);
```

```
   ba=b*a;
   aba=a*b*a;
   PRINT a b, e, h ba, aba;
QUIT;
```

Output – Example 4.12:

```
            a                       b
        1       2       1   -1.5      2  -1.5
        2       3       2    1       -1    1
        1       2       1    0        0    0

                        e
              1       0       1
              0       1       0
              0       0       0

        h                      ba
        1       0       1      1    0    1
        0       1       0      0    1    0
        0       0       0      0    0    0

                       aba
              1       2       1
              2       3       2
              1       2       1
```

The matrix **B** is the generalized inverse of the matrix **A**. The matrix, **E**, is the matrix, **A**, reduced to row-echelon normal form. When a matrix is reduced to row-echelon normal form, the rank of the matrix is equal to the number of rows consisting of non-zero elements. In this example, the rank of the matrix, **A**, is two. The matrix **H** is the Hermite form of the matrix **A**. When the matrix **A** is reduced to Hermite form, the rank of the matrix **A** is equal to the sum of the diagonal elements (the *trace* of the matrix) of the matrix **H**. Notice in the output that **BA=H** and **ABA=A**, confirming that **B** is the generalized inverse of **A**.

Example 4.13 – Repeat matrix and matrix of random numbers:

The REPEAT function can be used to create matrices by repeating arguments. This, like the DESIGN, DESIGNF, and others, is a method for producing a pattern matrix. The following code shows how to create patterned matrices with repeated blocks. This is useful for simulating distributions, as well as for design matrix applications.

```
PROC IML;
   n=5;
   m=3;
   mat = {1 2 3,4 5 6,7 8 9};
   rep = REPEAT(mat,m,n);
   norm = NORMAL(REPEAT(0,n,1));
   PRINT rep norm;
QUIT;
```

Output – Example 4.13:

```
      rep                                                    norm
      1  2  3  1  2  3  1  2  3  1  2  3  1  2  3              1
      4  5  6  4  5  6  4  5  6  4  5  6  4  5  6              1
      7  8  9  7  8  9  7  8  9  7  8  9  7  8  9             .3
      1  2  3  1  2  3  1  2  3  1  2  3  1  2  3              1
      4  5  6  4  5  6  4  5  6  4  5  6  4  5  6             -1
      7  8  9  7  8  9  7  8  9  7  8  9  7  8  9
      1  2  3  1  2  3  1  2  3  1  2  3  1  2  3
      4  5  6  4  5  6  4  5  6  4  5  6  4  5  6
      7  8  9  7  8  9  7  8  9  7  8  9  7  8  9
```

The statement

```
rep = REPEAT(mat,m,n);
```

creates a new matrix, **REP**, by stacking the matrix, **MAT**, m times down and n times across. Of use is the $n \times 1$ column vector **NORM**. It is created with the statement

```
norm = NORMAL(REPEAT(0,n,1));.
```

The elements of the vector are random numbers generated from a standard normal distribution using 0 as the seed. The NORMAL function is an IML function similar to the RANNOR function used in data processing steps. Generating vectors and matrices of random numbers from a specified distribution is useful when using IML for simulations.

4.4 Chapter Exercises

4.1 Use the following data to create the 3×4 matrix **A**. Create a row vector **COL_TOT_A** consisting of the column totals for the columns of the matrix **A**. Create a column vector **ROW_TOT_A** consisting of the row totals for the rows of the matrix **A**. Re-direct the output to a pdf document. Print the matrix **A**, the row vector **COL_TOT_A**, and the column vector **ROW_TOT_A** to the PDF document. The PDF document should also contain a descriptive title at the top of the document and descriptive labels (rather than the defaults) centered above each matrix and vector.

$$A = \begin{bmatrix} 1 & 3 & 5 & 7 \\ 2 & 4 & 6 & 8 \\ 9 & 0 & 9 & 0 \end{bmatrix}$$

4.2 Use the following matrix **A** to create a 2×2 submatrix, **B**, consisting of the 2,2; 2,3; 3,2; and 3,3 elements of the matrix **A**. Create a row vector **COL_TOT_B** consisting of the column totals for the columns of the matrix **B**. Create a column vector **ROW_TOT_B** consisting of the row totals for the rows of the matrix **B**. Re-direct the output to an HTML document. Print the matrix **B**, the row vector **COL_TOT_B**, and the column vector **ROW_TOT_B** to the HTML document. The HTML document should also contain a descriptive title at the top of the document and descriptive labels (rather than the defaults) centered above each matrix and vector

$$A = \begin{bmatrix} 1 & 3 & 5 & 7 \\ 2 & 4 & 6 & 8 \\ 9 & 0 & 9 & 0 \end{bmatrix}$$

4.3 Create a column vector, **C**, consisting of the diagonal elements of the matrix, **D**.

$$D = \begin{bmatrix} 1 & 2 & 3 \\ 2 & 3 & 4 \\ 3 & 4 & 5 \end{bmatrix}$$

4.4 A popular game involves a search to identify and sink ships located on a grid. Involved in the game is identifying the length and direction (horizontal or vertical) of the ships. Consider the situation in which the following grid is displayed:

$$\begin{bmatrix} 0 & 0 & 0 & 1 & 0 & 0 \\ 0 & 0 & 0 & 1 & 0 & 0 \\ 0 & 0 & 0 & 1 & 0 & 0 \\ 0 & 0 & 0 & 0 & 0 & 0 \\ 0 & 0 & 0 & 0 & 0 & 0 \\ 0 & 0 & 0 & 0 & 0 & 0 \end{bmatrix}$$

The zeros represent empty space and the ones represent the target. Write a module in IML that will accept as input the name of a matrix and output the location (row, column) of the target, including its starting location, ending location and its length. So, for the above grid/matrix, the output would be

Starting Point: (1,4) Ending Point: (3,4) Length: 3

Verify the accuracy of the module using the following matrices:

$$
\begin{bmatrix} 0\,0\,0\,0\,0\,0 \\ 0\,1\,1\,1\,1\,1 \\ 0\,0\,0\,0\,0\,0 \\ 0\,0\,0\,0\,0\,0 \\ 0\,0\,0\,0\,0\,0 \\ 0\,0\,0\,0\,0\,0 \end{bmatrix}
\begin{bmatrix} 0\,0\,0\,0\,0\,0 \\ 0\,0\,0\,0\,0\,0 \\ 0\,1\,1\,1\,0\,0 \\ 0\,0\,0\,0\,0\,0 \\ 0\,0\,0\,0\,0\,0 \\ 0\,0\,0\,0\,0\,0 \end{bmatrix}
\begin{bmatrix} 0\,0\,0\,0\,0\,0 \\ 0\,0\,0\,0\,0\,1 \\ 0\,0\,0\,0\,0\,1 \\ 0\,0\,0\,0\,0\,1 \\ 0\,0\,0\,0\,0\,1 \\ 0\,0\,0\,0\,0\,1 \end{bmatrix}
\begin{bmatrix} 1\,1\,1\,0\,0\,0 \\ 0\,0\,0\,0\,0\,0 \\ 0\,0\,0\,0\,0\,0 \\ 0\,0\,0\,0\,0\,0 \\ 0\,0\,0\,0\,0\,0 \\ 0\,0\,0\,0\,0\,0 \end{bmatrix}
$$

4.5 Use the direct product operator, @, the functions I and J, and the horizontal concatenation operator, ||, where appropriate to produce the following matrices:

$$
\begin{bmatrix} 1\,1\,1\,0\,0\,0 \\ 1\,1\,1\,0\,0\,0 \\ 1\,1\,1\,0\,0\,0 \\ 0\,0\,0\,1\,1\,1 \\ 0\,0\,0\,1\,1\,1 \\ 0\,0\,0\,1\,1\,1 \end{bmatrix}
\begin{bmatrix} 1\,1\,0\,0\,0\,0 \\ 1\,1\,0\,0\,0\,0 \\ 0\,0\,1\,1\,0\,0 \\ 0\,0\,1\,1\,0\,0 \\ 0\,0\,0\,0\,1\,1 \\ 0\,0\,0\,0\,1\,1 \end{bmatrix}
\begin{bmatrix} 1\,0\,0 \\ 1\,0\,0 \\ 0\,1\,0 \\ 0\,1\,0 \\ 0\,0\,1 \\ 0\,0\,1 \end{bmatrix}
\begin{bmatrix} 1\,1\,0\,1\,0\,0 \\ 1\,1\,0\,0\,1\,0 \\ 1\,1\,0\,0\,0\,1 \\ 1\,0\,1\,1\,0\,0 \\ 1\,0\,1\,0\,1\,0 \\ 1\,0\,1\,0\,0\,1 \end{bmatrix}
$$

4.6 Use the IML functions J and I, and the direct product operator, @, and the horizontal concatenation operator, ||, to create the following design matrix **X**. Within a PROC IML step create a SAS data set X with four variables from the four columns of the matrix **X**. specify the four variable names to be **C**, **A1**, **A2**, and **A3**. Print the SAS data set X to the output to verify its accuracy.

$$
\mathbf{X} = \begin{bmatrix} 1\,1\,0\,0 \\ 1\,1\,0\,0 \\ 1\,0\,1\,0 \\ 1\,0\,1\,0 \\ 1\,0\,0\,1 \\ 1\,0\,0\,1 \end{bmatrix}
$$

4.7 Create the IML matrix **X** from the previous exercise using a method other than that described in the previous exercise. Print the matrix **X**, including a descriptive title to a text file.

4.8 Create a module named STDEV that computes the standard deviations of each of the columns of an m×n matrix. Check it with the following matrix, **S**, to see if it returns the values 1, 2, 3, and 4.

$$
\mathbf{S} = \begin{bmatrix} 0\,3\,1\,1 \\ 1\,5\,4\,5 \\ 2\,7\,7\,9 \end{bmatrix}
$$

4.9 Create a module named COLTOT that gives column totals (sums) for the columns of a given matrix. Check it with the matrix, **S**, from Exercise 4.9 to see if it returns the values 3, 15, 12, and 15.

4.10 Create a module named RCTOTALS that produces two vectors, **R**, and **C**. The vector **R** should be a column vector consisting of the row totals of the input matrix. The vector **C** should be a row vector consisting of the column totals of the same input matrix. Test the module using the matrix **S** from Exercise 4.10 to see if it returns the following two vectors:

$$\mathbf{R} = \begin{bmatrix} 5 \\ 15 \\ 25 \end{bmatrix} \quad \mathbf{C} = \begin{bmatrix} 3 & 15 & 12 & 15 \end{bmatrix}$$

4.11 Create a module named **Z_VEC** that standardizes a column vector by subtracting the mean of the vector from each element in the vector, and dividing each resulting difference by the standard deviation of the vector. Test the module using the column vector, **R**, from Exercise 4.11 to see if it returns the vector

$$\mathbf{Z\_VEC} = \begin{bmatrix} -1 \\ 0 \\ 1 \end{bmatrix}.$$

4.12 Consider the experiment of rolling one or more six-sided fair dice and adding up the values of the dice as the result. Create a module named

```
DICE( cubes, num)
```

that accepts the inputs cubes, the number of dice; and num, the result of interest. The module then determines the number of possible ways the result, num, can be rolled by the dice. The module then prints a sentence indicating the number of cubes used, the result of interest, and the number of ways that result can occur. So, assuming the input values were cubes=2 and num=4, the output would read, "There are 2 ways the number 4 can occur by rolling 2 dice." This experiment counts the occurrence of the first cube reading a 1 and the second cube reading a 3 (1 and 3) the same as if the first cube reads a 3 and the second cube reads a 1 (3 and 1). Consequently, there are two ways to obtain a result of 4 by rolling two dice: (1 and 3) and (2 and 2). Test the module DICE using cubes=1, num=3 and cubes=4, num=7.

4.13 Repeat the previous problem but consider the occurrence of the first cube reading a 1 and the second cube reading a 3 (1 and 3) different from the occurrence of the first cube reading a 3 and the second cube reading a 1 (3 and 1). Test the module using cubes=1, num=3 and cubes=4, num=7.

Chapter 5
Mathematical and Statistical Basics

5.1 Transposes, Traces, and Ranks

There are many statistical equations that require finding the transpose of a matrix, the trace, rank, etc. Some of these operations are basic functions with sensible names that are easily discovered and used in IML. Some operations are less obvious.

5.1.1 Transpose

The transpose of a matrix can be computed in two different ways. One way is to use the function T. The statement

```
t_a = T(a);
```

will compute the transpose of the matrix **A**. The other method for finding the transpose of a matrix is using the tick mark (`). The statement

```
t_b = b';
```

will compute the transpose of the matrix **B**. The transpose of a matrix changes the order of the elements in a matrix such that the rows become columns and the columns become rows. Consider the following example:

Example 5.1 – Methods for computing the rank of a matrix:

The following code computes the transpose of the matrix **A**.

```
PROC IML;
    a={1 2 3,4 5 6};
    t_a=T(a);
    PRINT a t_a;
QUIT;
```

Output – Example 5.1:

	a			t_a	
1	2	3		1	4
4	5	6		2	5
				3	6

Notice that a 3×2 matrix (**T_A**) is computed from a 2×3 matrix (**A**). The two rows of **A** have become the two columns of **T_A**.

5.1.2 Trace

The trace of a matrix is found in a way one might expect. The statement

```
trace = TRACE(a);
```

will compute the trace of the matrix **A**, meaning the sum of the diagonal elements of the matrix **A**.

5.1.3 Rank

Finding the rank of a matrix using IML is not so obvious. The obvious way to find the rank of a matrix, **A**, would be to write out the function

```
rank = RANK(a); /**** does NOT compute the rank of the
   matrix A ****/.
```

However, the RANK function in IML does not find the rank of a matrix. The RANK function in IML creates a matrix of ranks based on the elements of a given matrix. So, one must be a little clever. A statement that can be used to find the rank of a matrix, **A**, using IML is

```
rank = ROUND(TRACE(GINV(a)*a));
```

This statement makes use of the fact that the rank of A is equal to the rank of AA^-, and since AA^- is idempotent, its rank is equal to its trace. For a proof, refer to Graybill, 2000.

It is of note that this method for finding the rank of the matrix **A** requires inverting the matrix **A**. If the matrix A is large, its inversion can be computational intensive. A more efficient way to find the rank of a matrix is to find it in such a way that does not require inverting the matrix. Consider the following example:

Example 5.1 – Methods for computing the rank of a matrix:

Three different methods for computing the rank of the matrix **A** are demonstrated here. The first method requires inversion of the matrix **A**. The second and third methods do not require the inversion of the matrix **A**. The second method relies

on the ECHELON function to find the rank. The third method computes the rank
by computing the trace of the Hermite reduction of the matrix **A**. If matrix **A** is
small, there is little difference among the three methods in the time it takes SAS
to compute the rank of the matrix **A**. However, if the matrix **A** has more than 1000
rows or columns, the time it takes to invert the matrix **A** becomes significant. The
difference between the second and third methods is negligible. The third method
only works for square matrices, a requirement of the TRACE function.

```
PROC IML;
    a={1 2 3,
        4 5 6,
        7 8 9};
    PRINT a;
    r1=ROUND(TRACE(GINV(a)*a));
    START myrank(mat);
        e=echelon(mat);
        e=(e^=0)[,+];
        myrank=(e^=0)[+,];
        RETURN(myrank);
    FINISH;
    r2=MYRANK(a);
    r3=TRACE(HERMITE(a));
    PRINT r1 r2 r3;
QUIT;
```

The modular function MYRANK has been created to compute the rank of an input
matrix using the second method, while not requiring the computation of the inverse
of the input matrix. The newly-created function is then used to compute the rank of
the matrix **A**.

Output – Example 5.1:

	a	
1	2	3
4	5	6
7	8	9
r1	r2	r3
2	2	2

If the input matrix is square and is not large, any of the three methods will work.
If the input matrix is not square but is large, the second method is recommended
because it is efficient and does not require a square matrix. If the input matrix is
square and large either the second or third method will work nicely, but the third
method requires fewer lines of code. If the matrix is not square and not large, either
of the first two methods will work, but the first method requires fewer lines of
code.

5.1.4 Determinant

The determinant of a square matrix is computed as follows:

```
Det_A = Det(a);
```

where **A** is a square matrix.

5.1.5 Inverses

Let **B** be an n × n square nonsingular matrix such that **AB**=**I** then **B** is the inverse of **A**. The inverse of a matrix is found by the following:

```
Inverse = INV(a);
```

The Moore-Penrose inverse (Graybill 2000, 2001), which will be referred to as the *MP-inverse*, does not have the property of the inverse. However, it does have the following four properties:

AB is symmetric idempotent
BA is symmetric idempotent
ABA = **A**
BAB = **B**

A restriction of the inverse is that it requires the matrix to be inverted be square. The MP-inverse does not have that restriction. The MP-inverse is found by the following:

```
mpinverse = GINV(a); .
```

Example 5.2 – The least squares estimator for σ^2:

This example employs the rank and MP-inverse in computing the least squares estimator for σ^2. The equation,

$$s^2 = \frac{y'(I_n - X(X'X)^- X')y}{n-p},$$

where $(X'X)^-$ represents the MP-inverse of $(X'X)$, can be entered directly in IML in the following manner:

```
PROC IML;
    x={1 2 2,2 3 4,4 3 8};
    y={-1,-2,-2};
    n=NROW(x);
    p=ROUND(TRACE(GINV(x)*x));
    in=I(n);
    s2=(1/(n-p))*y'*(in-x*GINV(x'*x)*x')*y;
    PRINT p s2;
QUIT;
```

Output – Example 5.2:

```
        p         s2
        2  0.0645161
```

Notice, the matrix **X** is of rank **P**=2 which is less than full rank (**N**=3).
The generalized inverse, also called the *conditional inverse*, has the following property:
The matrix B is a generalized inverse of A if and only if

ABA = A

There are potentially several generalized inverses for a given matrix. Also, all MP-inverses inverses are generalized inverses, though not all generalized inverses are MP-inverses. A generalized inverse satisfies only one of the properties of the MP-inverse (Graybill 2000, 2001). As with MP-inverses, generalized inverses can be used to solve systems of linear equations. Generalized inverses are easier to compute manually than MP-inverses. However, with the speed and facility of personal computers, the benefit of the generalized inverse over the MP-inverse ceases. See Example 4.12 for more information about the generalized inverse.

Note: Though the function GINV computes the generalized inverse of a matrix, the function CINV does not compute the conditional inverse of a matrix, but rather is used to compute a quantile from the Chi-square distribution.

5.2 Descriptive Statistics

Descriptive statistics are those basic summary measures that might be encountered in an introductory statistics class, or used in basic business reports or sporting events. Many of the examples in this section are more appropriately solved using other SAS procedures like MEANS, FREQ, or UNIVARIATE. However, if the researcher is using IML for other reasons and is already in the IML computing environment, these methods will, in most cases, be more efficient than leaving the IML computing environment, computing the statistics in a PROC UNIVARIATE step, and returning to the IML computing environment. These examples are also useful in academic settings. Descriptive statistics include (but are not limited to) the following:

mean, variance, frequency, percentile, quartile, sum, standard deviation

There is a method in IML for creating several descriptive (summary) statistics for SAS data sets using a SUMMARY statement. The SUMMARY statement has many features and options.

Example 5.3 – SUMMARY statistics:

The following code computes the following summary statistics:

count, sum, maximum, mean, variance, standard deviation

```
DATA outcomes;
   INPUT col1 col2 col3;
   DATALINES;
   1 2 3
   4 5 6
   7 8 9
   ;
RUN;

PROC IML;
   USE outcomes;
   SUMMARY VAR   {col1 col2 col3}
           STAT {N SUM MAX MEAN VAR STD}
           OPT  {SAVE};
   CLOSE outcomes;
   PRINT _NOBS_,col1,col2,col3;
QUIT;
```

The USE statement specifies which SAS data set to use in the SUMMARY statement. The VAR option indicates which variables from the SAS data set will be
used for computing the summary statistics. The STAT option indicates the summary
statistics to compute on each variable. The OPT option with the SAVE operand indicates that all of the summary statistics computed are to be saved in matrices. These
matrices are then printed.

Output – Example 5.3: Basic Statistics Resulting From the SUMMARY Statement

Nobs	Variable	N	SUM	MAX	MEAN	VAR	STD
3	COL1	3	12.00000	7.00000	4.00000	9.00000	3.00000
	COL2	3	15.00000	8.00000	5.00000	9.00000	3.00000
	COL3	3	18.00000	9.00000	6.00000	9.00000	3.00000

The above output of summary statistics prints by default. To suppress the printing
of the summary statistics in that default table, include the operand NOPRINT with
the OPT option.

Output – Example 5.3 (continued):

		NOBS			
		3			
		COL1			
3	12	7	4	9	3
		COL2			
3	15	8	5	9	3
		COL3			
3	18	9	6	9	3

The above output is a result of the PRINT statement.

The following are additional ways that IML can compute summary statistics. One purpose for including this section, in addition to the SUMMARY function, is to give alternative methods that might better suit the researcher's purposes, and because by modifying the code in the following examples summary statistics as well as modified summary statistic, not included with SUMMARY or other functions, can be computed.

5.2.1 Sum

Sums are desirable for row or column totals, and as an intermediate step for finding a mean or variance, matrix totals, etc. The following example demonstrates computations of sums using the SUM function and the sum operator.

Example 5.4 – Sums:

This example computes the following:

VEC_SUM: The sum of the elements of the first row of the matrix **X**
COL_TOT: The column totals (sum of the elements in each column of the matrix **X**
MAT_SUM: The sum of the elements of the matrix **X**
SUBMAT: A submatrix including the first two columns and first two rows of the matrix **X**
SM_SUM: The sum of the elements of a submatrix including the first two rows and the first two columns of the matrix **X**

```
PROC IML;
   x={1 2 3,4 5 6,7 8 9};
   vec_sum=x[1,+];
   col_tot=x[+,];
   mat_sum=SUM(x);
   submat=x[1:2,1:2];
```

```
    sm_sum=SUM(submat);
    PRINT vec_sum col_tot mat_sum sm_sum;
QUIT;
```

Output – Example 5.4

VEC_SUM	COL_TOT			MAT_SUM	SM_SUM
6	12	15	18	45	12

5.2.2 Mean

There are several methods for computing means that may be of interest: a vector mean, the mean of the elements of a matrix, the mean of the elements of a submatrix. The following example includes computations of each:

Example 5.5 – Means:

This example computes the following:

VEC_MEAN: The mean of the elements of the first row of the matrix **X**
MAT_MEAN: The mean of the elements of the matrix **X**
SM_MEAN: The mean of the elements of a submatrix including the first two rows and the first two columns of the matrix **X**

```
PROC IML;
    x={1 2 3,4 5 6,7 8 9};
    vec_mean=x[1,:];
    mat_mean=x[:];
    submat=x[1:2,1:2];
    sm_mean=submat[:];
    PRINT vec_mean mat_mean sm_mean;
QUIT;
```

Output – Example 5.5

VEC_MEAN	MAT_MEAN	SM_MEAN
2	5	3

5.2.3 Variance

There are several methods for computing variances that may be of interest: a vector variance, the variance of the elements of a matrix, the variance of the elements of a submatrix. The following example includes computations of each:

Example 5.6 – Variances:

This example computes the following:

VEC_VAR: The variance of the elements of the first row of the matrix **X**
MAT_VAR: The variance of the elements of the matrix **X**
SM_VAR: The variance of the elements of a submatrix including the first two
                  rows and the first two columns of the matrix **X**

```
PROC IML;
  x={1 2 3,4 5 6,7 8 9};
  ssq=SSQ(x[1,]);
  sum=x[1,+];
  vec_var=(ssq-sum*sum/NCOL(x))/(NCOL(x)-1);
  jn=J(NROW(x));
  mat_mean=x[:]));
  mat_var=SSQ(x-mat_mean*jn)/(NROW(x)*NCOL(x)-1);
  submat=x[1:2,1:2];
  ssq_sub=SSQ(submat);
  sum_sub=SUM(submat);
  sm_var=(ssq_sub - sum_sub*sum_sub/(NCOL(submat)
     *NROW(submat))) /
  (NCOL(submat)*NROW(submat)-1);
  PRINT vec_var mat_var sm_var;
QUIT;
```

Output – Example 5.6

VEC_VAR	MAT_VAR	SM_VAR
1	7.5	3.3333333

5.2.4 Other Basic Statistics

Several basic statistics (count, sum, median, min, max, quartiles, etc.) are found by simply applying an IML function or the mathematical operator equivalent. Others (frequency, mode, quantile, etc.) involve more work. However, modules can be written so that even the more computationally-involved statistics can be found by simply applying a module call statement. The following example computes several additional basic statistics, some from IML functions, some formulaically, and some using modules.

Example 5.7 – Other basic statistics:

This example computes the following: count, sum, mean, median, mode, variance, standard deviation, frequency, quartiles, percentile, and quantile. The code is broken up into sections to facilitate the description of each section of code.

```
PROC IML;
   scores={90,62,66,68,70,72,73,74,78,78,78,79,80,81,82,
           82,82,84,84,85,85,85,85,87,88,89,89,89,89,61};
   PRINT scores;
   count=NROW(scores);
   sum=SUM(scores);
   mean=sum/count; /* =scores[:] */
   median=MEDIAN(scores);
```

The above section of code begins the PROC IML step. The data are entered through manual entry and the column vector **SCORES** is printed to the output. The **COUNT** statistic is computed using the NROW function. The **SUM** statistic is computed using the SUM function. The **MEAN** statistic is computed using a formulaic approach (for academic purposes only) by dividing the **SUM** by the **COUNT**. The **MEDIAN** statistic is computed using the MEDIAN function.

```
/** Frequency Distribution **/
START freq(x);
   unique=UNION(x)`;  /** vector of unique scores **/
   observed=J(NROW(unique),1);
   DO i=1 TO NROW(unique);
      observed[i]= NCOL(LOC(x=unique[i]));
   END;
   freq=unique||observed;  /* vector of unique scores and
      their frequencies */
   f_Label={"    SCORE" "FREQUENCY"}; /* labels for the
      columns of Freq */
    PRINT "Frequency Distribution",freq [COLNAME=f_label];
FINISH;
CALL FREQ(scores);
/** End Frequency Distribution**/
```

The above code is used to create the module FREQ for the purpose of computing a frequency distribution for the data. The vector **X** is passed to the module in the START statement. The UNION function is designed to create a vector of unique numbers from two or more input vectors. The resulting vector consists of unique numbers that exist in each of the original vectors. The numbers in the resulting vector are listed in ascending order. However, if a single input vector is passed as an argument to the UNION function, the result is a vector consisting of unique numbers from the input vector, listed in ascending order–just what is needed here. The unique values of the input vector are assigned to the column vector **UNIQUE**. The vector **OBSERVED** is created and will contain the frequency for each unique value in the **UNIQUE** vector. The statement

```
        observed[i]= NCOL(LOC(x=unique[i]));
```

counts the number (NCOL) of locations (LOC) in which a unique value (**UNIQUE**) resides in the input vector, and populates the elements of the column vector **OBSERVED** with those frequencies. The CALL statement then calls the newly-created module FREQ and passes **SCORES** as the input vector. The PRINT statement prints the frequency distribution to the output. Though no output matrices are created as a result of this module, the code can be modified to produce **FREQ** as an output matrix if desired.

```
/** Start Mode **/
START mode(x);
    unique=UNION(x)';
    observed=J(NROW(unique),1);
    DO i=1 TO NROW(unique);
        observed[i]= NCOL(LOC(x=unique[i]));
    END;
    max=MAX(observed);
    mode=unique[LOC(observed=max),1];
    RETURN(mode);
FINISH;
mode=MODE(scores);
/** End Mode **/
```

The above code is used to create a modular function that will compute the mode(s) of the input vector. The column vectors **UNIQUE** and **OBSERVED** are created within the module because their previous definitions in the code are not passed to the MODE module and hence are not accessible within the MODE module. The statement

```
    max=MAX(observed);
```

identifies the highest frequency of occurrence of any value in the input data. The mode(s) will have to have a frequency of occurrence equal to **MAX** to be considered the mode(s). The statement

```
    mode=unique[LOC(observed=max),1];
```

identifies the values within the column vector **UNIQUE** corresponding to the frequencies in the column vector **OBSERVED** that are equal to the maximum observed frequency, **MAX**.

```
ssq=SSQ(scores);
variance=(ssq-sum*sum/count)/(count-1);
stdev=SQRT(variance);
quartiles=QUARTILE(scores);
```

The above code computes the variance (formulaically for academic purposes), the standard deviation, and the quartiles for the input vector **SCORES**. The SSQ function is used here only as an intermediate step to computing the variance and standard deviation.

```
/** Start Percentile **/
START percentile(x,pct);
    CALL SORT(x,1);
    count=NROW(x);
    pct=pct/100*(count+1);
    fpct=FLOOR(pct);
    cpct=CEIL(pct);
    IF fpct=cpct THEN percentile=x[pct];
    ELSE percentile=x[fpct]+(pct-fpct)* (x[cpct]-x[fpct]);
     /* SAS PCTLDF 4 */
    RETURN(percentile);
FINISH;
percentile=PERCENTILE(scores,75);
/** End Percentile **/
```

The above code creates a modular function that will compute a percentile based on an input column vector and the numeric percentage value of interest. The statement

```
CALL SORT(x,1);
```

sorts the input vector in ascending order. The data must be sorted before a percentile can be computed. The original input **PCT** value is an integer (i.e.- to compute the 75th percentile, the integer 75 is assigned to the input variable **PCT**). The computation of the percentile is based upon definition 4 in the UNIVARIATE procedure of SAS. The statement

```
pct=pct/100*(count+1);
```

assigns the location of the input percentage in the input column vector to the variable **PCT**, overwriting its input value. If the new value for **PCT** is an integer, that integer identifies the row in the input vector associated with the percentile. If **PCT** is not an integer, the percentile is a computed value. The percentile value is assigned to the variable **PERCENTILE**, and returned by the modular function.

```
/** End Percentile **/
CALL SORT(scores,1); /* sorts the scores in ascending
    order */
quantile=scores[CEIL(count*.45)];
PRINT count sum mean median mode,
      variance stdev quartiles, "75th Percentile:"
      percentile, "45th Quantile:" quantile;
QUIT;
```

The above code sorts the values within the input vector in ascending order. This is necessary for the subsequent computation of the quantile of interest. The computation of the quantile is based upon definition 3 of a percentile in the UNIVARIATE procedure of SAS. The statement

```
quantile=scores[CEIL(count*.45)];
```

computes the 45th quantile (due to the .45 in the equation) of the input column vector **SCORES**. The PRINT statement then prints computed statistics to the output.

Output – Example 5.7

```
                              scores
                                 90
                                 62
                                 66
                                 68
                                 70
                                 72
                                 73
                                 74
                                 78
                                 78
                                 78
                                 79
                                 80
                                 81
                                 82
                                 82
                                 82
                                 84
                                 84
                                 85
                                 85
                                 85
                                 85
                                 87
                                 88
                                 89
                                 89
                                 89
                                 89
                                 61
```

Frequency Distribution
freq

SCORE	FREQUENCY
61	1
62	1
66	1
68	1
70	1

72	1
73	1
74	1
78	3
79	1
80	1
81	1
82	3
84	2
85	4
87	1
88	1
89	4
90	1

count	sum	mean	median	mode
30	2395	79.833333	82	85
				89

variance	stdev	quartiles
67.867816	8.2381925	61
		74
		82
		85
		90

	percentile
75th Percentile:	85.5

	quantile
45th Quantile:	81

5.3 Chapter Exercises

5.1 Describe the differences among the inverse, generalized inverse, and MP-inverse.

5.2 Which of the following statements are always true?

    a. If **B** is a generalized inverse of **A** then **B** is an MP-inverse of **A**.
    b. If **B** is the inverse of **A** then **B** is an MP-inverse of **A**.
    c. If **B** is the MP-inverse of **A** then **B** is an inverse of **A**.
    d. If **B** is the MP-inverse of **A** then **B** is a generalized inverse of **A**.

5.3 Which of the following statements can sometimes be true? For statements that are sometimes true, use IML to give an example to show that it is sometimes true.

   a. If **B** is a generalized inverse of **A** then **B** is an MP-inverse of **A**.
   b. If **B** is the inverse of **A** then **B** is an MP-inverse of **A**.
   c. If **B** is the MP-inverse of **A** then **B** is an inverse of **A**.
   d. If **B** is the MP-inverse of **A** then **B** is a generalized inverse of **A**.

5.4 Modify the following code to compare the speed of the different methods of
     computing the rank of the matrix **A**. If a difference in speed is not noticeable,
     increase the value of the variable **DIM** until it is. Note: If you receive an error
     message related to memory, your computer may not have sufficient memory to
     compute the rank of a matrix the size of **A** using all of the different methods.

```
PROC IML;
   Dim=1000
   a=NORMAL(REPEAT(0,dim,dim));
   rank1=ROUND(TRACE(a*GINV(a)));
   PRINT rank1;
QUIT;
```

5.5 Using IML to compute a rank, find the dimension of the column space of **B**,
     where

$$\mathbf{B} = \begin{bmatrix} 1 & 2 & 3 \\ 2 & 1 & 2 \\ 3 & 1 & 1 \\ -1 & 0 & 2 \end{bmatrix}$$

5.6 For the 4×1 column vector **V**,

$$\mathbf{V} = \begin{bmatrix} 1 \\ 2 \\ -1 \\ 3 \end{bmatrix}$$

   show using IML that

$$\mathbf{MP\_V} = \frac{\mathbf{V}'}{\mathbf{V}'\mathbf{V}},$$

   where **MP_V** is the MP-inverse of **V**. It can be shown that this is true of any
   non-zero $n \times 1$ column matrix **V**. Using IML, come up with two more examples
   like **V** that support this claim.

5.7 Using IML, show which of the following matrices are full column rank:

$$\mathbf{A} = \begin{bmatrix} 9 & 7 & 9 \\ 4 & 2 & 3 \\ 8 & 0 & 1 \\ 6 & 9 & 0 \end{bmatrix} \quad \mathbf{B} = \begin{bmatrix} 1 & 2 & 3 \\ 2 & 1 & 2 \\ 3 & 1 & 1 \\ -1 & 0 & 2 \end{bmatrix} \quad \mathbf{C} = \begin{bmatrix} 1 & 0 & 1 \\ 0 & 1 & 1 \\ 1 & 1 & 1 \\ 1 & 1 & 0 \end{bmatrix}$$

Chapter 6
Linear Algebra

6.1 Matrix Algebra

IML automates intensive matrix algebraic statements. Problems that could take hours to solve on paper can take seconds to solve using IML. Conventional mathematical notation is used as much as possible for simplicity, so the way a problem would be written down on a piece of paper is similar to how it would be written in IML code. The following examples incorporate many of the mathematical operators and functions that IML provides.

Example 6.1 – Linear equations: the least squares estimator of β*, where* X *is full-column rank.*

The equation, $b = (X'X)^{-1} X'y$, can be entered directly in IML code. Here is an example of the code that could be used to compute this value:

```
PROC IML;
    x={1 2,2 3,4 3};
    y={-1,-2,-2};
    b=INV(x`*x)*x`*y;
    PRINT b;
QUIT;
```

The INV function computes the inverse of the matrix in the parentheses. The tic mark (`) transposes the matrix to its immediate left – the function T also computes the transpose of a matrix. The asterisk (*) performs standard multiplication of the immediate matrices on either side.

Output – Example 6.1:

```
          B
    -0.096774
    -0.548387
```

J.J. Perrett, *A SAS/IML Companion for Linear Models*, Statistics and Computing,
DOI 10.1007/978-1-4419-5557-9_6, © Springer Science+Business Media, LLC 2010

Example 6.2 – Quadratic equations: the least squares estimator for σ^2 and σ.

The equation,

$$s^2 = \frac{y'(I_n - X(X'X)^- X')y}{n - p},$$

where $(X'X)^-$ represents the MP-inverse of $(X'X)$, can be entered directly in IML code*. Here is an example of the code that could be used:

```
PROC IML;
    x={1 2,2 3,4 3};
    y={-1,-2,-2};
    n=NROW(x);
    p=ROUND(TRACE(GINV(x)*x));
    in=I(n);
    s2=(1/(n-p))*y'*(in-x*GINV(x'*x)*x')*y;
    s=SQRT(s2);
    PRINT s2 s;
QUIT;
```

The function NROW is used to compute the sample size, **N**. The scalar **P** is assigned the rank of the matrix **X**. The matrix **IN** is defined as an $n \times n$ identity matrix. The backslash (/) performs standard scalar division. The GINV function computes the MP-inverse of $X'X$. The SQRT function is used to compute the standard deviation by computing the square root of the variance.

Output – Example 6.2:

```
            S2              S
       0.0645161 0.2540003
```

6.2 Matrix Decomposition

The decomposition of a matrix into a product of matrices of various forms and properties has many uses in the analysis of linear models. IML matrix decomposition functionality includes the following:

- APPCORT Call: orthogonal decomposition
- COMPORT Call: orthogonal decomposition (Householder)

---

*Note: Entering the equation for a variance directly is intended for academic purposes (i.e.- seeing how the theoretical equation works in practice). In rare data situations, Alternative algorithms are more numerically accurate. For example, the variance formula used to compute MAT_VAR on page 99 is preferred over the variance formula used to compute SM_VAR on that same page (See Thisted 1988).

- GSORTH Call: Gram-Schmidt orthonormalization
- HALF Function: Cholesky decomposition (identical to ROOT Function)
- QR Call: QR decomposition (Householder)
- ROOT Function: Cholesky decomposition (identical to HALF Function)
- SVD Call: Singular value decomposition

The following two examples demonstrate the use of SVD Call and GSORTH Call. Example 9.3 gives an example of the ROOT Function.

Example 6.3 – Singular value decomposition (Moore-Penrose inverse):

The SVD call performs singular value decomposition on an $m \times n$ matrix, **A**, such that

$$A = U^* diag(Q)^* V'$$

where $U'U = V'V = I_n$, $UU' = I_m$, and Q contains the singular values of A. The form of the statement is:

Call SVD(u, q, v, a);

This call is subject to the following condition: $m = n$.

One use of matrix decomposition is to compute the Moore-Penrose inverse of a matrix. The following code demonstrates the singular value decomposition of the matrix **A**, the properties of the resulting matrices, and the use of this decomposition to compute the MP-inverse.

```
Proc IML;
    a = {1 2 3 4,
         5 6 7 8,
         9 0 1 2};
    RESET FUZZ;
    CALL SVD(u,q,v,a);
    upu = u'*u;
    vpv = v'*v;
    vvp = v*v';
    uup = u*u';
    a2 = u*DIAG(q)*v';
    ginva=GINV(a);
    m = NROW(a); n = NCOL(a);
    DO i=1 TO n;
        IF q[i] <= 1E-12 * q[1] then q[i] = 0;
        ELSE q[i] = 1 / q[i];
    END;
    ga = v*DIAG(q)*u';
    PRINT a a2„ u q v„ upu uup„ vpv vvp„ ginva ga;
QUIT;
```

Output – Example 6.3

A				A2			
1	2	3	4	1	2	3	4
5	6	7	8	5	6	7	8
9	0	1	2	9	0	1	2

	U				Q	
0.3282727	-0.272589	0.9043962		0	0.0651364
0.842727	-0.347968	-0.410767		0	0.1366617
0.4266712	0.8970027	0.1154895		0	1.1465517
					0

	V		
0.5459703	0.8282512	-0.12616	0
0.3721185	-0.359828	-0.751921	-0.408248
0.4761851	-0.322048	-0.053535	0.8164966
0.5802517	-0.284269	0.6448509	-0.408248

	UPU					UUP	
1	0	0	0	1	0	0
0	1	0	0	0	1	0
0	0	1	0	0	0	1
0	0	0	0			

	VPV					VVP		
1	0	0	0	1	0	0	0
0	1	0	0	0	1	0	0
0	0	1	0	0	0	1	0
0	0	0	1	0	0	0	1

	GINVA				GA	
-0.15	0.05	0.1	-0.15	0.05	0.1
-0.758333	0.3916667	-0.133333	-0.758333	0.3916667	-0.133333
-0.033333	0.0666667	-0.033333	-0.033333	0.0666667	-0.033333
0.6916667	-0.258333	0.0666667	0.6916667	-0.258333	0.0666667

Note that **A2** = **A** and that **GA** = **GINVA** (the MP-inverse of **A**).

Example 6.4 – Gram-Schmidt orthonormalization:

The GSORTH call computes the Gram-Schmidt orthonormalization. The form of the statement is:

Call GSORTH(*p*, *t*, *lindep*, *a*);

Given an m × n input matrix **A**, the GSORTH call computes an m × n column-orthonormal matrix, **P**, an n × n upper-triangular matrix, **T**, and a scalar, *lindep*, indicating whether the columns of **A** are independent or dependent (*lindep*=0 if independent, *lindep*=1 if dependent). Note the following:

1. $\mathbf{B} = \mathbf{P}^*\mathbf{T} = \mathbf{A}$
2. *lindep* = 0 indicates that the columns of **A** are linearly independent
3. $\mathbf{P}^{-1} = \mathbf{P}'$
4. **T** is nonsingular

Although Note 1 always holds, Notes 2 through 4 occur when the columns of **A** are linearly independent.

The matrices **P** and **T** are used in many applications of linear models (see Graybill, 2000)

```
PROC IML;
   a={1 2 3,2 3 5,3 7 11};
   CALL GSORTH(p,t,lindep,a);
   p_inv=INV(p);
   b=p*t;
   PRINT a b,p p_inv,t lindep;
QUIT;
```

Output – Example 6.4:

A			B		
1	2	3	1	2	3
2	3	5	2	3	5
3	7	11	3	7	11

P			P_INV		
0.2672612	-0.051434	-0.96225	0.2672612	0.5345225	0.8017837
0.5345225	-0.822951	0.1924501	-0.051434	-0.822951	0.5657789
0.8017837	0.5657789	0.1924501	-0.96225	0.1924501	0.1924501

T			LINDEP
3.7416574	7.750576	12.294017	0
0	1.3887301	1.9545091	
0	0	0.1924501	

6.3 Solving Systems of Linear Equations

IML has a few different functions designed for solving systems of linear equations, including the following:

- HOMOGEN Function: Solves a homogenous system of linear equations
- SOLVE Function: Solves a linear system of equations
- TRISOLV Function: Solves a linear system of equations using triangular matrices

The following two examples are two of them.

Example 6.5 – Solving AX=0 for X:

The HOMOGEN function solves homogeneous systems of equations of the form **AX**=0 for **X**, where **A** is an $n \times p$ matrix of rank $r < p$, and **X** is a $p \times 1$ vector. The result of the HOMOGEN function is a matrix, **X**, which contains $p-r$ columns of orthogonal vectors which each satisfies the condition $A^*X[,I]=0$ (where $X[,I]$ is the ith column vector of the matrix X).

```
PROC IML;
   RESET FUZZ;
   a={10 5 15, 12 6 18, 14 7 21, 16 8 24};
   x=HOMOGEN(a);
   zero1=a*x[,1];
   zero2=a*x[,2];
   PRINT a x zero1 zero2;
QUIT;
```

Output – Example 6.5:

a			x		zero1	zero2
10	5	15	0	0.8451543	0	0
12	6	18	0.9486833	-0.169031	0	0
14	7	21	-0.316228	-0.507093	0	0
16	8	24			0	0

The matrix, A, is a 4×3 matrix of rank $r = 1$. The matrix, X, contains two column vectors that will satisfy the condition $A^*X[,I]=0$ (where $X[,I]$ is the ith column vector of the matrix X). The 4×1 column vectors, **ZERO1** and **ZERO2** verify that the column vectors, $X[,1]$ and $X[,2]$ of the matrix X do, in fact, satisfy the condition $A^*X[,I]=0$. In the code, the

```
RESET FUZZ;
```

statement formats the output such that any numeric result with numbers smaller than 1E-12 (default) is set to zero. The FUZZ option default of 1E-12 can be altered by assigning a value (ex. RESET FUZZ=1E-15). That creates cleaner output since often with computations a very small number will result due to finite precision arithmetic. If the

```
RESET FUZZ;
```

statement is not used, small numbers will appear in scientific notation. Because in this example the vectors **ZERO1** and **ZERO2** are expected to contain zeros, it is appropriate to use the

```
RESET FUZZ;
```

statement in this instance.

Example 6.6 – Solving AX=B for X:

Given an $n \times n$ non-singular matrix A and an $n \times p$ matrix B, the $n \times p$ matrix X that solves the equation $AX=B$ is computed in the following manner:

```
PROC IML;
    a={1 2 3,6 5 4,0 7 8};
    b={1, 2, 3};
    x=SOLVE(a,b);
    ax=a*x;
    PRINT a x b ax ;
QUIT;
```

Output – Example 6.6:

a			x	b	ax
1	2	3	0.0238095	1	1
6	5	4	0.2380952	2	2
0	7	8	0.1666667	3	3

The solution is fundamentally equivalent to using the statement

`x=INV(a)*b;`

However, the SOLVE function is numerically more stable and more accurate – and is therefore recommended over using

`x=INV(a)*b;`

When working with a singular matrix **A**, the above formula can be replaced with

`x=GINV(a)*b;`

which will use the MP-inverse instead of the inverse. The solution, **X**, will still be unique.

The matrix, **AX**, verifies that the matrix, **B**, satisfies the equation **AX=B**.

6.4 Eigenvalues and Eigenvectors

The SAS/IML product has three commands used for working with eigenvalues and eigenvectors (also referred to as characteristic roots and characteristic vectors):

- EIGEN Call   Computes the eigenvalues and eigenvectors for square matrix
- EIGVAL Function   Computes the eigenvalues for a square matrix
- EIGVEC Function   Computes the eigenvectors for a square matrix
- GENEIG Call   Computes the eigenvalues and eigenvectors for a generalized eigenproblem

Among the many uses of eigenvalues and eigenvectors in linear models applications are the following:

- Verifying orthogonality of vectors
- Determining the existence of a solution vector in a linear system
- Determining if a square matrix is singular or nonsingular
- Determining if a symmetric matrix is definite or semidefinite in nature
- Determining whether or not a symmetric matrix is idempotent
- Computing the rank or trace of a matrix

The following examples demonstrate basic uses of eigenvalues and eigenvectors.

Example 6.7 – Definite nature of a symmetric matrix using eigenvalues:

To determine properties of the $n \times n$ matrix **A**, the eigenvalues are computed. Although the EIGVAL Function is used in this example, the EIGEN call could also have been used to find the eigenvalues.

```
Proc IML;
a={ 1   0   1,
    0   4  -1,
    1  -1   2};
   RESET FUZZ;
   eigval=EIGVAL(a);
   rank=ROUND(TRACE(GINV(a)*a)); /** rank of matrix A **/
   PRINT a , ,eigval, ,rank;
QUIT;
```

Output – Example 6.7:

a

1	0	1
0	4	-1
1	-1	2

eigval

4.4605049
2.2391233
0.3003719

rank

3

This example demonstrates the following about **A**:
   Because the eigenvalues are all > 0 for the symmetric matrix **A**,

1. **A** is positive definite (Graybill, 2000).
2. The rank of **A** is $n=3$.
3. **A** is nonsingular.

Example 6.8 – Eigenvalues and eigenvectors:

Let **A** be an n × n matrix, **X** be an n × 1 eigenvector of **A**, λ be an eigenvalue of **A**. Then

$$AX = \lambda X$$

This relationship can be demonstrated with the following code:

```
Proc IML;
A={ 1  0  1,
    0  4 -1,
    1 -1  2};
  RESET FUZZ;
  CALL EIGEN(eigval,eigvec,a);
  ax1=a*eigvec[,1];
  ax2=a*eigvec[,2];
  ax3=a*eigvec[,3];
  lx1=eigval[1]*eigvec[,1];
  lx2=eigval[2]*eigvec[,2];
  lx3=eigval[3]*eigvec[,3];
  PRINT a,eigval,eigvec,ax1 lx1,ax2 lx2,ax3 lx3;
QUIT;
```

Although the EIGEN call was used in the program, the EIGVAL and EIGVEC functions could also have been used.

Output – Example 6.8:

```
                            a
          1            0            1
          0            4           -1
          1           -1            2

                         eigval
                       4.4605049
                       2.2391233
                       0.3003719

                         eigvec
          -0.12 0.5744266 0.8097123
      0.9017526 0.4042222 -0.153123
      -0.415261 0.7117854 -0.566498

                    ax1          lx1
          -0.535262 -0.535262
          4.0222721 4.0222721
          -1.852276 -1.852276
```

```
       ax2          1x2

   1.286212   1.286212
   0.9051033  0.9051033
   1.5937753  1.5937753

       aXx          1x3

   0.2432148  0.2432148
  -0.045994  -0.045994
  -0.17016   -0.17016
```

The vector EIGVAL contains the three eigenvalues of **A**. The matrix **EIGVEC** contains three eigenvectors of **A** associated with the three eigenvalues.

The output shows that indeed the equation $\mathbf{AX}=\lambda\mathbf{X}$ holds true for all three eigenvectors and eigenvalues.

6.5 Chapter Exercises

6.1 Find examples of at least three different statistical analyses presented in elementary-level statistics textbooks. Duplicate all of the computations using IML.

6.2 Many basic statistics can be computed in IML using either functions or operators. For example, the arithmetic sum of the elements of the matrix **B** can be found using both of the following methods:

```
a.  sum=b[+];
b.  sum=SUM(b);
```

6.3 List two different ways of computing the following statistics: mean, standard deviation, range, minimum, maximum, median. Test your methods on the following matrix **B**.

$$\mathbf{B} = \begin{bmatrix} 1\,2\,3\,4 \\ 2\,3\,4\,5 \\ 3\,4\,5\,6 \end{bmatrix}$$

6.4 Using IML functions, find a matrix **P** such that

$$\mathbf{P'P=A},$$

where

$$\mathbf{A} = \begin{bmatrix} 1\,2\,4 \\ 2\,3\,5 \\ 4\,5\,6 \end{bmatrix}.$$

Describe the IML function that was used, its purpose, and how it solved this problem.

6.5 Find the eigenvalues of the matrix **C**, where

$$\mathbf{C} = \begin{bmatrix} 1 & 0 & 1 \\ 2 & 1 & 1 \\ 1 & 1 & 0 \end{bmatrix}.$$

Explain the relationship between the rank of the matrix **C** and its eigenvalues.

6.6 Describe matrix decomposition and its purpose. Give a specific example using IML.

6.7 Describe the differences among the IML functions HOMOGEN, SOLVE, and TRISOLV. Give one example of each function using different data for each. Explain why the linear system is more appropriately solved using the function chosen rather than one of the other two.

Chapter 7
The Multivariate Normal Distribution

7.1 Multivariate Normal Random Variable

Matrices simplify computations involving the multivariate normal distribution. Suppose the random variable X is a $p \times 1$ normally distributed random variable, $X \sim N(\mu, \Sigma)$, where μ is a $p \times 1$ vector of means and Σ is a $p \times p$ covariance matrix. Now, suppose $X = \begin{bmatrix} X_1 & X_2 & \cdots & X_p \end{bmatrix}$ is an $n \times p$ data matrix. Consider X as a matrix containing p samples, each of size n. Estimates of μ and Σ may be computed using the following formulas:

$$\hat{\mu} = \left(\frac{1}{n}\right) j_n'X \qquad (7.1)$$

$$\hat{\Sigma} = \left(\frac{1}{n-1}\right) X'\left(I_n - \frac{1}{n}J_n\right)X \qquad (7.2)$$

where I_n is an $n \times n$ identity matrix, j_n is an $n \times 1$ vector of 1's, and J_n is an $n \times n$ matrix of 1's.

Example 7.1 – Parameter estimates of multivariate normally distributed data:

Let $X = \begin{bmatrix} x1 \\ x2 \end{bmatrix}$ represent a bivariate normal random variable, $X \sim N(\mu, \Sigma)$, where mean $\mu = \begin{bmatrix} \mu_1 \\ \mu_2 \end{bmatrix}$ and covariance $\Sigma = \begin{bmatrix} \sigma_{11} & \sigma_{12} \\ \sigma_{21} & \sigma_{22} \end{bmatrix}$. A sample of size $n=10$ is generated as follows.

The following code computes the estimate of the mean vector and covariance matrix based on Equations 7.1 and 7.2.

```
PROC IML;
   x={4.4 5.5,2.4 3.1,5.5 6.1,7.6 6.3,7.4 7.5,
      8.5 10.2,0.6 1.5,4.5 4.1,7.2 5.8,2.8 2.5};
   n=NROW(x);
   jn=J(n,1);
   jnn=J(n,n,1);
```

J.J. Perrett, *A SAS/IML Companion for Linear Models*, Statistics and Computing, DOI 10.1007/978-1-4419-5557-9_7, © Springer Science+Business Media, LLC 2010

```
    in=I(n);
    mean=(1/n)*(jn'*x)';
    cov=(1/(n-1))*(x'*(in-(1/n)*jnn)*x);
    PRINT mean cov;
QUIT;
```

Table 7.1 Bivariate Normal Data

x1	x2
4.4	5.5
2.4	3.1
5.5	6.1
7.6	6.3
7.4	7.5
8.5	10.2
0.6	1.5
4.5	4.1
7.2	5.8
2.8	2.5

Output – Example 7.1:

```
      mean          cov

      5.09  6.8165556  6.1828889
      5.26  6.1828889  6.5915556
```

7.2 Linear Function of a Multivariate Normal Random Variable

Now, consider a linear function of a multivariate normally distributed random variable. Suppose the random variable X is a $p \times 1$ normally distributed random variable, $X \sim N(\mu, \Sigma)$. Suppose A is an $m \times p$ matrix of constants and b is a $m \times 1$ vector of constants. Then,

$$AX + b \sim N(A\mu + b, A\Sigma A'). \tag{7.3}$$

Example 7.2 – Parameter estimates of a linear function of multivariate normal random variables:

In this example, let A be a (5×2) matrix of constants, B be a (2×1) column vector of constants, and X be a (1000×5) matrix of values generated from a Normal distribution with a mean of zero and a variance of one. Let **Y=A*X+B**. The following code computes the estimate of the mean vector and covariance matrix of **Y**, using Equations 7.1 and 7.2.

```
PROC IML;
    a={-2 -1 0 1 2,2 -1 -2 -1 2};
    x=NORMAL(REPEAT(0,1000,5));
    b={1,2};
    y=a*x'+b;
    n=NROW(x);
    jn=J(n,1);
    jnn=J(n,n,1);
    in=I(n);
    mean_x=(1/n)*(jn'*x);
    cov_x=(1/(n-1))*(x'*(in-(1/n)*jnn)*x);
    mean1_y=a*mean_x'+b;
    cov1_y=a*cov_x*a';
    mean2_y=    (1/n)*(jn'*y')';
    cov2_y= (1/(n-1))*(y*(in-(1/n)*jnn)*y');
    PRINT mean_x[FORMAT=5.2], cov_x[FORMAT=5.2],
          mean1_y[FORMAT=5.2], cov1_y[FORMAT=5.2],
          mean2_y[FORMAT=5.2], cov2_y[FORMAT=5.2];
QUIT;
```

Output – Example 7.2:

```
                         mean_x
          0.03   0.05   -0.00   0.01   0.01
                         cov_x
          1.08   0.01  -0.02 -0.06  -0.00
          0.01   1.04  -0.02  0.06  -0.01
         -0.02  -0.02   1.01  0.03   0.03
         -0.06   0.06   0.03  0.95  -0.02
         -0.00  -0.01   0.03 -0.02   0.95

                        mean1_y
                         0.91
                         2.01
                        cov1_y
                    10.22 -0.95
                    -0.95 14.51
                        mean2_y

                         0.91
                         2.01
```

```
cov2_y
10.22 -0.95
-0.95 14.51
```

The matrix **X** is generated from a normally-distributed random variable with mean of zero and variance of one. The values in **MEAN_X** in the output are close to the mean of zero, and the values in **COV_X** in the output are close to one on the diagonals and close to zero on the off-diagonals—what we would expect. These values get closer to their expected values as the size of the generated sample (1000 in this case) increases. The 2×1000 matrix **Y** represents transformed variables that are a function of **X**. Specifically, two random variables with means associated with the values in the column vector **B** and covariance matrix associated with the values in the matrix **A**. The values of MEAN1_Y=MEAN2_Y and COV1_Y=COV2_Y demonstrate that $\hat{\mu}$ and $\hat{\Sigma}$ from Equations 7.1 and 7.2 can be found using Equations 7.1 and 7.2 on the transformed data **Y**, as well as by using the formulas

$$\hat{\mu}_Y = A\hat{\mu}_X + b$$

and

$$\hat{\Sigma}_Y = A\hat{\Sigma}_X A'$$

Based upon the formulas in Equation 7.3. The mean values of Y, **MEAN1_Y** and **MEAN2_Y**, have values of 0.91 and 2.01 that are close to the values of **A**, 1 and 2, respectively. The diagonal elements of **COV1_Y** and **COV2_Y** are close to the expected variances 10 and 14, found by squaring and summing the elements of each row of **A**, and multiplying the result by the value of **COV_X**. Linear functions of this type are used in the estimation and testing of linear contrast in ANOVA settings.

7.3 Quadratic Function of a Multivariate Normal Random Variable

Next, consider a quadratic function of a multivariate normally distributed random variable. Suppose the random variable Y is a $p \times 1$ normally distributed random variable, $Y \sim N(\mu, \Sigma)$ where Σ has rank p. Suppose A is a $p \times p$ matrix of constants. Then, $U = Y'AY \sim \chi^2(p, \lambda)$, a non-central chi-squared distribution with p degrees of freedom and non-centrality parameter $\lambda = \frac{1}{2}\mu'A\mu$ if and only if $A\Sigma$ is an idempotent matrix of rank p; or ΣA is an idempotent matrix of rank p (i.e.- $\Sigma A(\Sigma A) = \Sigma A$); or Σ is a conditional inverse of A (i.e.- $A\Sigma A = A$) and A has rank p. Obviously, if $\mu = 0$ then $\lambda = 0$ and U becomes a central chi-square random variable.

Because SAS is designed to work with data rather than analytic problems, the random variables described in this section need data evaluations to be considered by SAS. Whereas each of the p values in the $p \times 1$ column vector Y are random variables, replacing them with a constant value would not be conducive to this illustration as the mean of each constant value would be the value itself and the variance of each constant value would be zero in all cases. Therefore, a sampling situation will be considered to illustrate the concepts presented in this section.

Let $\mathbf{Y}$ be a $p \times n$ matrix of values such that each row represents a $1 \times n$ row vector of observations generated from Normal random variable. As the sample size, n, gets larger, the $p \times 1$ sample mean vector and $p \times p$ sample covariance matrix approach the values μ and Σ. Consequently, $U = Y'AY$ approaches $\chi^2(p, \lambda)$.

Example 7.3 – Parameter estimates of a quadratic function of multivariate normal random variables:

The elements of the $p \times n$ matrix $\mathbf{Y}$ are generated from a standard normal distribution. The row means and the covariance matrix are computed and shown to be similar to a $p \times 1$ column vector of ones and a $p \times p$ identity matrix, respectively. A $p \times p$ matrix of constants, $\mathbf{A}$, is computed such that the conditions that Σ has rank p and $A\Sigma$ is an idempotent matrix of rank p are satisfied. The $n \times n$ matrix $\mathbf{U}$ is computed as

$$U = Y'AY$$

and then rank, mean, and covariance are computed to verify the results are those expected based upon the analytic equations.

```
PROC IML;
    p=5;
    n=3000;
    y=NORMAL(REPEAT(0,p,n));
    jn=J(n,1);
    jnn=J(n,n,1);
    in=I(n);
    mean_y=(1/n)*(y*jn);
    cov_y=(1/(n-1))*(y*(in-(1/n)*jnn)*y');
    rank_y=ROUND(TRACE(y*GINV(y)));
    rank_cov_y=ROUND(TRACE(cov_y*GINV(cov_y)));
    PRINT mean_y cov_y rank_y rank_cov_y;
    a=GINV(cov_y);
    rank_a=TRACE(HERMITE(a));
    u=y'*a*y;
    a_sigma=a*cov_y;
    a_sigma_a=a_sigma*a;
    PRINT a a_sigma_a;
    a_sigma_a_sigma=a_sigma*a_sigma;
    PRINT a_sigma a_sigma_a_sigma;
    lambda=.5*mean_y'*a*mean_y;
    rank_u=TRACE(HERMITE(u));
```

```
    mean_u=TRACE(a_sigma)+mean_y'*a*mean_y;
    cov_u=2*TRACE(a_sigma*a_sigma)+4*mean_y'*a_sigma*a*mean_y;
    PRINT rank_y lambda rank_a rank_u;
QUIT;
```

In this example, a sample of size n=3000 is used. Depending on the computer used, a larger sample size may prevent certain computations from working due to available memory requirements.

Output – Example 7.3:

```
    mean_y  cov_y                                  rank_y  rank_cov_y
      0.04    1.02 0.02 0.01     0 0.01              5           5
     -.01    0.02 1.01
     -.03    0.01 0.02
     -.01    0.01
     -.03    0.98 0.01     0
      0.01       0 0.01 0.01     1 0.01
      0.01    0.01 0.02     0 0.01 0.97
        a                            a_sigma_a
   0.98 -.02 -.01     0 -.01        0.98 -.02 -.01     0 -.01
   -.02 0.99 0.03 -.01 -.02        -.02 0.99 0.03 -.01 -.02
   -.01 0.03 1.02 -.01     0       -.01 0.03 1.02 -.01     0
      0 -.01 -.01     1 -.01           0 -.01 -.01     1 -.01
   -.01 -.02     0 -.01 1.03        -.01 -.02     0 -.01 1.03
      a_sigma                      a_sigma_a_sigma
     1    0    0    0    0          1    0    0    0    0
     0    1    0    0    0          0    1    0    0    0
     0    0    1    0    0          0    0    1    0    0
     0    0    0    1    0          0    0    0    1    0
     0    0    0    0    1          0    0    0    0    1
                    mean_u      cov_u     rank_u
                   5.00238   10.00952        5
```

The elements of the mean vector, **MEAN_Y**, are all close to zero. The larger the sample size, **N**, used in the program, the closer these values will be to zero. The matrix **COV_Y** is similar to the identity matrix. Both **Y** and **COV_Y** are of rank **P=5**. The matrix **A_SIGMA** is an idempotent matrix of rank **P=5**, **A_SIGMA_A** = **A**, and has rank **P=5**. Consequently, **U** has a central chi-squared distribution with **P=5** degrees of freedom (because the non-centrality parameter **LAMBDA=0**).

7.4 Independence of Linear and Quadratic Functions

Independence of linear and quadratic functions is important in determining distributions. Though this is generally an issue of theory and proof related to random variables rather than numeric application, a numeric example can be instructive.

Suppose the random variable Y is a $p \times 1$ normally distributed random variable, $Y \sim N(\mu, \Sigma)$ where S has rank p. If $A\Sigma B = 0$ then the following hold true: $Y'AY$ and $Y'BY$ are independent, AY and $Y'BY$ are independent.

Example 7.4 – Independence of linear and quadratic functions:

The following code computes the Covariance matrix for data generated from a Normal distribution. The independent 2×2 matrices **A** and **B** are then created and it is demonstrated that $A\Sigma B = 0$.

```
PROC IML;
    p=2;
    n=6000;
    y=NORMAL(REPEAT(0,p,n));
    jn=J(n,1);
    jnn=J(n,n,1);
    in=I(n);
    cov_y=(1/(n-1))*(y*(in-(1/n)*jnn)*y');
    a={1 0,
       0 0};
    b={0 0,
       0 1};
    a_sigma_b=a*cov_y*b;
    PRINT a, b, a_sigma_b;
QUIT;
```

Output – Example 7.4:

```
                        a

            1           0
            0           0

                        b

            0           0
            0           1

                  a_sigma_b

            0           0
            0           0
```

The sample size, **N**=6000, is large so that the sample covariance closely approximates the actual covariance matrix it was generated from. Smaller-sized generated samples, and even **N**=6000 will yield **A_SIGMA_B** matrices with the elements not exactly equal to zero.

7.5 Expected Value and Covariance of Linear and Quadratic Functions

Suppose the random variable X is a $p \times 1$ has expected value of μ and covariance Σ of rank p. The linear function $Y = AX + b$ has expected value $E(Y) = A\mu + b$ and covariance $cov(Y) = A\Sigma A'$. The quadratic function $U = X'AX$ has expected value $E(U) = tr(A\Sigma) + \mu'A\mu$ and covariance $cov(U) = 2tr(A\Sigma A\Sigma) + 4\mu'A\Sigma A\mu$, though the covariance formula holds only when X is normally distributed and A is symmetric.

Example 7.5 –Expected value and covariance of linear and quadratic functions:

This example uses information from Example 7.3 to find the mean and variance of the matrix **U** where

$$U = Y'AY$$

by applying data to the analytic formulas. The matrix **U** has a central chi-square distribution with degrees of freedom **P**=5. It can be shown that the mean of a chi-square random variable is equal to its degrees of freedom and that the variance is equal to twice the mean. In this example, the mean is found to be **MEAN_U**=5 and the variance is found to be **COV_U**=10 as expected. Though the computed values of the mean and variance are approximate, using a large sample size, **N**=6000, and the

```
RESET FW=4 SPACES=2;
```

results in the output values appearing to be appear exact.

```
options ls=75;
PROC IML;
   RESET FW=4 SPACES=2;
   p=5;
   n=6000;
   y=NORMAL(REPEAT(0,p,n));
   jn=J(n,1);
   jnn=J(n,n,1);
   in=I(n);
   mean_y=(1/n)*(y*jn);
   cov_y=(1/(n-1))*(y*(in-(1/n)*jnn)*y');
   a=GINV(cov_y);
   a_sigma=a*cov_y;
   a_sigma_a_sigma=a_sigma*a_sigma;
   mean_u=TRACE(a_sigma)+mean_y'*a*mean_y;
   cov_u=2*TRACE(a_sigma*a_sigma)+4*mean_y'
```

```
      *a_sigma*a*mean_y;
   print mean_u cov_u;QUIT;
```

Output – Example 7.5:

	mean_u	cov_u
	5	10

7.6 Chapter Exercises

7.1 Using IML, generate a 100×2 matrix containing two independent column vectors of random values from a normal distribution with mean 5 and variance 4 for the first column and mean 3 and variance 9 for the second column. Compute the vector containing the column means and the 2×2 covariance matrix. Comment on the computed values as they apply to the known means, variances, and covariances.

7.2 Consider the following matrices:

$$A = \begin{bmatrix} 2 & 9 \\ 3 & 6 \\ 7 & 3 \\ 4 & 9 \\ 9 & 3 \end{bmatrix}, B = \begin{bmatrix} 3 \\ 5 \end{bmatrix}, C = \begin{bmatrix} -0.30 \\ -1.28 \\ 0.24 \\ 1.28 \\ 1.20 \end{bmatrix},$$

Where **A** is a matrix of constants, **B** is a column vector of constants, and **C** is a vector of sample values generated from a normal distribution with a mean of zero and a variance of one. Let X be a 5×1 column vector of normally-distributed random variables with a mean vector of zeros and an identity covariance matrix.

a. Let $W=A*X+B$. According to theory, what is the expected mean vector and covariance matrix of W?
b. Let $Z=A*C+B$. Thus **Z** is a sample version of W. Using IML, compute **Z**, the estimate of the expected mean vector of W.
c. Compare the answers (the mean vectors) for Parts a and b.
d. Why can the covariance matrix of W not be estimated using the data given?

7.3 Let X be a 5×1 column vector of normally-distributed random variables with a mean vector of zeros and an identity covariance matrix. Let A be a 5×5 matrix of constants,

$$A = \begin{bmatrix} 1\ 2\ 3\ 5\ 7 \\ 2\ 3\ 5\ 7\ 5 \\ 3\ 5\ 7\ 5\ 3 \\ 5\ 7\ 5\ 3\ 2 \\ 7\ 5\ 3\ 2\ 1 \end{bmatrix}$$

a. Describe the theoretical distribution of $U = X'AX$.
b. Consider the column vector **C**,

$$C = \begin{bmatrix} -0.30 \\ -1.28 \\ 0.24 \\ 1.28 \\ 1.20 \end{bmatrix}$$

Using IML, compute the estimated mean and variance of $V=C'^*A^*C$.

c. Compare the results of Parts a and b.

Chapter 8
The General Linear Model

8.1 Linear Models

The subject of linear models, and all of its intricacies is too vast to be properly dealt with in a book of this nature. Instead, a few examples of general applications of linear models in IML will be demonstrated. These examples can then be modified according to the needs of the researcher.

8.2 Creating a Design Matrix

Many designed experiments use the general linear model

$$Y = X\beta + \varepsilon, \tag{8.1}$$

where X is an $n \times p$ matrix of known fixed numbers representing different experimental categories and levels. Let β be a $p \times 1$ column vector of unknown parameters. Let ε be an $n \times 1$ column vector of random errors. In such a case, the matrix, X, is called a "design matrix." IML has functions designed to make creation of design matrices faster, easier, and more efficient. The following three functions will be used:

@ – the direct product
DESIGN – the design function
DESIGNF – the full-rank design function

The direct product (also known as the Kronecker product) is an expansion operator. In mathematics, a direct product is written in the form A ⊗ B. In IML code, it is written in the form **A@B**. The result of the direct product A ⊗ B is a new matrix that multiplies each element of A with the entire matrix $\mathbf{B}^{\Psi}$. For example,

---

[Ψ] It should be noted that in some textbooks the result of the direct product $A \otimes B$ is a new matrix that multiplies each element of B with the entire matrix A – different from the way it is used in SAS/IML.

J.J. Perrett, *A SAS/IML Companion for Linear Models*, Statistics and Computing, 129
DOI 10.1007/978-1-4419-5557-9_8, © Springer Science+Business Media, LLC 2010

$$A = \begin{bmatrix} 2 & 0 \\ 1 & 2 \end{bmatrix}, \quad B = \begin{bmatrix} 1 & 2 & 3 \end{bmatrix}, \quad A \otimes B = \begin{bmatrix} 2 & 4 & 6 & 0 & 0 & 0 \\ 1 & 2 & 3 & 2 & 4 & 6 \end{bmatrix}.$$

The DESIGN and DESIGNF functions create design and full-rank design matrices, respectively, based on an input vector. The full-rank matrix is created by excluding the last column of the matrix that would have been created by using the same input vector in the DESIGN function. That last column is then subtracted from the other columns. It is sometimes referred to as "effect coding." The following example demonstrates the use of the DESIGN function.

Example 8.1 – Creating a design matrix:

A researcher wishes to analyze the model

$$Y_{ij} = \mu + \alpha_i + \tau_j + \varepsilon_{ij},$$

where i=1, 2, 3 and j=1, 2. This model equation can be rewritten as

$$\begin{bmatrix} Y_{11} \\ Y_{12} \\ Y_{21} \\ Y_{22} \\ Y_{31} \\ Y_{32} \end{bmatrix} = \begin{bmatrix} 1 & 1 & 0 & 0 & 1 & 0 \\ 1 & 1 & 0 & 0 & 0 & 1 \\ 1 & 0 & 1 & 0 & 1 & 0 \\ 1 & 0 & 1 & 0 & 0 & 1 \\ 1 & 0 & 0 & 1 & 1 & 0 \\ 1 & 0 & 0 & 1 & 0 & 1 \end{bmatrix} \times \begin{bmatrix} \mu \\ \alpha_1 \\ \alpha_2 \\ \alpha_3 \\ \tau_1 \\ \tau_2 \end{bmatrix} + \varepsilon.$$

The matrix of ones and zeros, the design matrix, can be created in IML in various ways. The following code will compute the design matrix three different ways:

```
PROC IML;
    a=3; t=2;
    y={5,10,3,6,3,9};
    x1={1 1 0 0 1 0,1 1 0 0 0 1,1 0 1 0 1 0,
        1 0 1 0 0 1,1 0 0 1 1 0,1 0 0 1 0 1};
    x2=J(a,1)@J(t,1)||I(a)@J(t,1)||J(a,1)@I(t);
    x3=DESIGN({1,1,1,1,1,1})||DESIGN({1,1,2,2,3,3})
        ||DESIGN({1,2,1,2,1,2});
    PRINT x1,x2,x3;
QUIT;
```

Output – Example 8.1: The Design Matrix

x1

1	1	0	0	1	0
1	1	0	0	0	1
1	0	1	0	1	0
1	0	1	0	0	1
1	0	0	1	1	0
1	0	0	1	0	1

x2

1	1	0	0	1	0
1	1	0	0	0	1
1	0	1	0	1	0
1	0	1	0	0	1
1	0	0	1	1	0
1	0	0	1	0	1

x3

1	1	0	0	1	0
1	1	0	0	0	1
1	0	1	0	1	0
1	0	1	0	0	1
1	0	0	1	1	0
1	0	0	1	0	1

8.3 Point and Interval Estimation

There are two common methods for solving a system of linear model equations to find the estimates of the unknown parameters: Least squares (LS) estimation and maximum likelihood (ML) estimation. Both methods require different assumptions and consequently different advantages (i.e.- if study parameters can't satisfy the assumptions of one method, perhaps they will satisfy the assumptions of the other method).

8.3.1 Least Squares Estimation

Consider the general linear model as defined in Section 8.2 (Equation 8.1) with the additional definition of ε having expected value $E[\varepsilon]=0$ (an $n \times 1$ vector of zeros) and covariance matrix $var[\varepsilon]=\sigma^2 I_n$ the identity matrix multiplied by an unknown scalar parameter. With this basic assumption about the random error associated with the model, Least Squares (LS) estimation allows for solving the system of linear model equations for estimates of the unknown parameters β and σ^2. The estimates are found to be

$$\hat{\beta} = (X'X)^{-1}X'Y$$

and

$$\hat{\sigma}^2 = \left(\frac{1}{n-p}\right) Y'(I_n - X(X'X)^{-1}X')Y,$$

which can be simplified to

$$\hat{\beta} = X^- Y \tag{8.2}$$

and

$$\hat{\sigma}^2 = \left(\frac{1}{n-p}\right) Y'(I_n - XX^-)Y, \tag{8.3}$$

where **N** represents the number of observations and **P** represents the rank of the matrix **X**. Notice that Equations 8.2 and 8.3 are written with MP-inverses rather than inverses. This allows for the flexibility of a singular **X** matrix. These estimators have very good properties with minimal assumptions. Notice there is no assumption that the random errors have a normal distribution. Though, if it does, the above estimators have even better properties.

Example 8.2 – LS estimates of β and σ²:

A researcher wishes to compute LS estimates for the parameters β and β^2 for the model

$$Y = X\beta + \varepsilon.$$

This model equation can be used to describe many different research settings. Consider the following setup:

$$
\begin{array}{ccc}
Y = & X & \beta + \varepsilon \\
\begin{bmatrix} Y_{11} \\ Y_{12} \\ Y_{21} \\ Y_{22} \\ Y_{31} \\ Y_{32} \end{bmatrix} = &
\begin{bmatrix} 1\,1\,0\,0\,1\,0 \\ 1\,1\,0\,0\,0\,1 \\ 1\,0\,1\,0\,1\,0 \\ 1\,0\,1\,0\,0\,1 \\ 1\,0\,0\,1\,1\,0 \\ 1\,0\,0\,1\,0\,1 \end{bmatrix} \times &
\begin{bmatrix} \mu \\ \alpha_1 \\ \alpha_2 \\ \alpha_3 \\ \tau_1 \\ \tau_2 \end{bmatrix} + \varepsilon.
\end{array}
$$

The following code can be used to compute the LS estimates defined in Equations 8.1 and 8.2.

```
PROC IML;
    y={5,10,3,6,3,9};
    x=DESIGN({1,1,1,1,1,1})||DESIGN({1,1,2,2,3,3})
        ||DESIGN({1,2,1,2,1,2});
    n=NROW(y);
    p=TRACE(HERMITE(x));  /** p=rank of x **/
```

```
   betas=GINV(x)*y;
   s2=(1/(n-p))*y'*(i(n)-x*GINV(x))*y;
   PRINT betas s2;
QUIT;
```

Output – Example 8.2:

```
          betas      s2
          3.27     1.17
          2.59
          -.41
          1.09
         -0.7
          3.97
```

8.3.2 Maximum Likelihood Estimation

One way of finding estimates for the unknown parameters in the general linear model is to find the values of β and σ^2 that maximize the likelihood function involving these unknown parameters and the data collected for the study.

Again consider the general linear model as defined in Section 8.3.1 with the additional assumption that the errors (ε) be normally distributed. The likelihood equation has the following form:

$$L(\mu, \sigma^2 : y) = (2\pi\sigma^2)^{-n/2} \, exp\left[\left(-\frac{1}{2\sigma^2}\right) \sum_{i=1}^{n} (y_i - \bar{y})^2 - \frac{n}{2\sigma^2}(\bar{y} - \mu)^2\right]$$

Often the log of this likelihood function is used rather than the likelihood function itself because the log of the likelihood function is mathematically easier to work with and maximizing one with respect to the unknown parameters achieves the same estimates as are achieved if maximizing the other. The log of the likelihood function has the following form:

$$logL(\mu, \sigma^2 : y) = -\frac{n}{2} log(2\pi\sigma^2) - \frac{1}{2\sigma^2} \sum_{i=1}^{n} (y_i - \bar{y})^2 - \frac{n}{2\sigma^2}(\bar{y} - \mu)^2 \quad (8.2)$$

There are various ways of finding values for the parameters that maximize the log of the likelihood equation. One method is to take the derivative(s) of the log of the likelihood equation with respect to each of the parameters, set the derivative(s) equal to zero and solve simultaneously for the pair of estimates. The second derivative(s) can then be used to determine if the obtained estimates identify a maximum, minimum, or saddle point.

Another method would be to plot the log of the likelihood function using a range of possible values in place of each of the parameters to identify the maximum of the function and the values of the parameters that achieve that maximum. Yet another method of finding the values for the parameters that maximize the log of the likelihood function is to use a numerical method in the form of a nonlinear optimization routine. There are a number of different nonlinear optimization routines available in SAS/IML.

Consider the following examples:

Example 8.3 – Maximum likelihood estimation (mathematical formulas):

By taking the derivatives of the log of the likelihood function (Equation 8.2) with respect to each of the unknown parameters, setting the equations equal to zero, and then solving them simultaneously it can be shown that the ML estimates are

$$\tilde{\mu} = \bar{y} = \frac{\sum_{i=1}^{n} y_i}{n}$$

and

$$\tilde{\sigma}^2 = \frac{\sum_{i=1}^{n} (y_i - \bar{y})^2}{n}$$

Note that the ML estimate of the variance $\left(\tilde{\sigma}^2\right)$ is not identical to its LS estimate $\left(\hat{\sigma}^2\right)$, but that $\tilde{\sigma}^2 = (n-1)\hat{\sigma}^2/n$. Examples 8.4 and 8.5 demonstrate applications of maximum likelihood estimation.

Example 8.4 – Maximum likelihood estimation (manual grid search):

The following code provides a module, norm_mle, that uses a manual grid search to find the maximum of the log of the likelihood function associated with the Normal distribution with unknown parameters being the mean and variance, and thereby determine approximations to the ML estimates of the two parameters. The module requires that the column vector containing the data be identified. The module can be modified to include more flexibility by allowing the user to pass additional values such as minimum, maximum, and increment values to the module. The purpose is to show that approximations to the maximum likelihood estimates can be found using a grid search. Pseudo random data values are generated to demonstrate the process. Values at equal intervals form a grid pattern of possible mean and variance values (the two unknown parameters). The log of the likelihood function is evaluated for each set of grid values. The resulting log-likelihood evaluations are considered and the maximum found. The values of the mean and variance that result in the maximum of the log-likelihood are then output as the "maximum likelihood estimates." These values can be compared to the vector mean and variances (the ML estimate for the variance is expected to differ from the sample variance computed using the SUMMARY function in SAS/IML as SUMMARY gives a LS estimate rather than an ML estimate). The log-likelihood values are then plotted to show the pattern of the function and the peak at the maximum. The grid could continually be redefined at smaller intervals to provide more precise estimates.

The DATA step in the first section of the code creates the SAS data set, dist that will contain the variable y. The DATA step populates the variable y with 100 pseudo-random values from a Normal distribution with mean=4 and standard deviation=3, and.

Next, in the PROC IML step, the data are read into a 100×1 column vector y. The SUMMARY function is used to display the vector mean and variance for comparison with the approximations to the ML estimates found later in the program.

The START command starts the module NORM_MLE that is intended to compute the approximations to the maximum likelihood estimates for the unknown parameters of the Normal density. The values **N**, **PI**, **YBAR**, and **SSQ** are defined. They will be used in the computation of the log-likelihood. The variables **MU_GRID** and **S2_GRID** specify the parameters of the grid search involving differing levels of **MU** and **S2**, respectively. The variable **GRID_N** is computed as the total number of combinations of **MU** and **S2** that will be used for the grid search and its computation is based upon the parameters of the grid search found in **MU_GRID** and **S2_GRID**. The grid will contain combinations of MU and S2 with values from 3 to 5 in increments of .1 and values from 5 to 15 in increments of .1, respectively.

The matrix **ITER** is created to hold all iterations of the grid search. The values are all initialized to 999. Those values will be replaced with each iteration of the current grid values and the evaluation of the log-likelihood at the grid values. The matrix is defined to have **GRID_N** rows and three columns. The variable **COUNT** follows the iterations to index the rows of the **ITER** matrix and is incremented at each step in the iteration.

The variable, **LOGLIK** is set to the evaluation of the log of the likelihood function with the value of the mean and variance being the current grid values. The log of the likelihood function is Equation 9.1 in computational form. A row of the **ITER** matrix is then populated with the current grid and log-likelihood values.

The definition of the 1×3 row vector, **MAX**, uses both the MAX function and the LOC function. The MAX function identifies the maximum value of the log-likelihood from among all the log-likelihood values computed during the grid search. LOC function then identifies the location of the maximum and the row vector containing the maximum log-likelihood value is assigned to **MAX**. That vector will contain the values of the mean and variance that obtained the maximum of all the computed log-likelihood values to a certain degree of precision. The row vector, **MAX**, is then printed to the output for the user to compare to the estimates computed by the SUMMARY function.

Next, a SAS data set, **LIKOUT**, is created containing the grid and log-likelihood values. The FINISH command marks the finish of the module NORM_MLE. Finally, PROC G3D is used to create a 3-dimensional wireframe plot of the log-likelihood function for the given values of the mean and variance, with a title created using the TITLE statement. The user can then view the plot of the log of the likelihood function and observe the maximum of the function as the highest point in the graph. That highest point is associated with the values determined to be the approximations of the ML estimates. The precision of these estimates can be improved by using smaller increments in the grid.

```
/** Generate 100 values from a Normal dist with
      mean=4 and std dev=3 **/
DATA dist;
   DO i=1 TO 100;
      y=RANNOR(0)*3+4;
      OUTPUT;
   END;
RUN;
PROC IML;
   USE dist;
   READ ALL VAR {y} INTO y;
   SUMMARY VAR {y} STAT {MEAN VAR};
   /** View mean and variance **/
   START norm_mle(y); /** Start norm_mle module **/
      n=NROW(y);
      pi = CONSTANT("PI");
      ybar=y[:];
      ssq=SSQ(y-J(n,1,1)*ybar);
      /** Specify parameters of grid search **/
      mu_grid={3,5,.1};
      s2_grid={5,15,.1};
      grid_n=((mu_grid[2]-mu_grid[1])/mu_grid[3]+1)
         *((s2_grid[2]-s2_grid[1])/s2_grid[3]+1);
      /** Initialize vector to hold grid values and
         log-likelihood **/
      iter=J(grid_n,3,999);
      count=0;
      DO mu=mu_grid[1] TO mu_grid[2] BY mu_grid[3];
      /** Grid values for mu **/
         DO s2=s2_grid[1] TO s2_grid[2] BY s2_grid[3];
         /** Grid values for variance **/
            loglik=(-n/2)*LOG(2*pi*s2)-(1/(2*s2))
            *ssq-(n/(2*s2))*(ybar-mu)*(ybar-mu);
            /** Computation of log-likelihood using
               current grid values **/
            count=count+1;
      iter[count,1]=mu;
      iter[count,2]=s2;
      iter[count,3]=loglik;
         END;
      END;
      max=iter[LOC(iter[,3]=MAX(iter[,3])),];
      /** Identify grid values associated with maximum
         of log-likelihood **/
```

```
   PRINT max [COLNAME={mu s2 loglik} FORMAT=8.1];
   CREATE likout FROM iter [COLNAME={mu s2 loglik}];
   /** Create output data set with grid and
       log-likelihood values **/
   APPEND FROM iter;
FINISH; /** Finish module norm_mle **/
RUN norm_mle(y);   /** Run module norm_mle **/
QUIT;


TITLE "Log Likelihood Based on Mean=4, Variance=9";
PROC G3D DATA=likout;
   PLOT mu*s2=loglik;
   /** Wireframe plot likelihood function at grid values **/
RUN;
```

Output – Example 8.4:

```
              Analysis variable : y

          Nobs          MEAN             VAR
      ---------------------------------------------
          100        4.03449          8.58704
      ---------------------------------------------
```

The above output results from the SUMMARY function, containing the sample mean and variance (LS estimates). By multiplying the variance by $(n - 1)/n = .99$ the LS variance estimate is converted to a value similar to that expected for the approximation to the ML variance estimate, 8.5011696.

Output – Example 8.4 (continued):

```
                              max
                     MU        S2     LOGLIK
                     4.0       8.5    -248.9
```

The above output is the 1×3 row vector, **MAX**, containing the maximum value of the computed log-likelihoods along with the values of the mean and variance used to obtain the maximum value. These mean and variance values would be considered the approximations to the ML estimates using the grid search method. As can be seen, the approximations are similar to the LS estimates in the previous portion of output. Accurate to within one number to the right of the decimal, the precision on these approximations can be improved by reducing the size of the intervals between numbers in the grid.

Log Likelihood Based on Mean=4, Variance=9

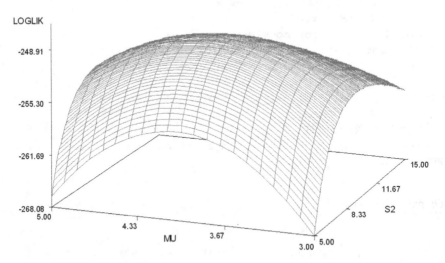

Fig. 8.1   3-D plot of log likelihood

Output – Example 8.4 (continued):

The above output is the plot produced by the PROC G3D step and shows the log of the likelihood function based on grid values for the mean and variance. The highest point in the plane is indexed by the approximations to the ML estimates for the mean and the variance.

Example 8.5 – Maximum likelihood estimation (nonlinear optimization routine):

A nonlinear optimization routine can be used to identify the ML estimates for the unknown parameters in a likelihood function. The following code uses a nonlinear optimization routine to maximize the likelihood function and thereby determine the ML estimates (or approximations of such estimates) of the two parameters.

```
/** Generate 100 values from a Normal dist.
    with mean=4 and std dev=3 **/
DATA dist;
  DO i=1 TO 100;
    y=RANNOR(0)*3+4;
    OUTPUT;
  END;
RUN;
PROC IML;
  USE dist;
```

```
READ ALL VAR {y} INTO y;
SUMMARY VAR {y} STAT {MEAN VAR};
/** View mean and variance **/
n=NROW(y);
pi = CONSTANT("PI");
ybar=y[:];
ssq=SSQ(y-J(n,1,1)*ybar);
START reml(theta) GLOBAL(n,pi,ssq,ybar);
   mu = theta[1];
   s2 = theta[2];
func = (-n/2)*LOG(2*pi*s2)-(1/(2*s2))*ssq-(n/(2*s2))
   *(ybar-mu)*(ybar-mu);
   RETURN(func);
FINISH reml;
x0 = {2. 15.}; /** Initial values **/
optn = {1 1};
/** First option indicates maximization, second
   option indicates amount of output **/
con={. 0}; /** Constraint that variance > 0 **/
CALL NLPNRA(rc,covs,"reml",x0,optn,con);
f_opt = reml(covs); /** Evaluation of final likelihood **/
PRINT "Reason for termination of optimization:"
   rc [FORMAT=2.0 LABEL=""],,,
   "ML Estimates:"
   covs [COLNAME={Mean Variance} LABEL=""],,,
   "Maximum of Log-likelihood
   Function:" f_opt [LABEL=""];
QUIT;
```

The value for **PI** (i.e.- 3.14...) is assigned using the CONSTANT function. The value for **YBAR**, the sample mean, is computed using the mathematical mean operator (:) designed to find the arithmetic average of all elements in a given matrix.

As an intermediate step, the sum of squared deviations is computed prior to defining the log-likelihood. This is done to reduce the number of computations performed in the optimization routine. It can well be noted that the values defined prior to the log-likelihood function are the sufficient statistics— Σy through the mean, **YBAR**, and $\Sigma(y^2)$ through the sum of the squared deviations, **SSQ**. Because they do not depend on the value of the unknown parameters, they can be evaluated from the data prior to assembling the log-likelihood function. In that way they are evaluated only once and not for each iteration of the optimization routine.

Modular coding is then used to create the function REML. In this case, a function is a type of module that returns a value in an assignment statement. Consequently, the value of the log-likelihood is computed within the module, REML, and later assigned to the variable **F_OPT**. The RETURN statement indicates that the value of the log-likelihood, **FUNC**, is to be the value returned by the function REML.

The 1×2 row vector **THETA** is to be the variable input for the function. The values **N**, **PI**, **SSQ**, and **YBAR** are values that are needed for the log-likelihood equation. Outside the module, **N**, **PI**, **SSQ**, and **YBAR** are globally defined (defined on the global symbol table). They are passed into the module REML (transferred to the local symbol table for the module) using the GLOBAL option; otherwise, only variables defined within the module would be recognized within the module (see Section 3.1.2 for more information about symbol tables).

Next, the values **X0**, **OPTN**, and **CON** are created. The variable **X0** is defined as a 1×2 row vector containing the starting values of the unknown mean and variance for the optimization routine. The values 2 and 5 were chosen for illustration. In practice, the researcher would decide the initial values thought to be close to the actual parameter values. The closer the initial values to the actual unknown parameter values, the more likely the optimization routine is to find the maximum of the log-likelihood function. If the log-likelihood function is multi-modal, poor choices of initial values could lead the optimization routine to find a local maximum rather than the overall maximum.

The variable **OPTN** contains options for the optimization routine. The first number in the 1×2 row vector refers to whether or not a maximum or a minimum is sought. Because the log-likelihood is to be maximized, the number one is used. A zero would request the optimization routine minimize the function. The second number in the 1×2 row vector specifies the amount of printed output to be produced by the optimization routine. The value 1 specifies the printing of the iteration history as well as summaries for the start and termination of the optimization. A zero for this second number (the default) would specify that no output is to be produced.

The variable **CON** represents constraints on the parameters. The 1×2 row vector contains a decimal for the first constraint implying no constraint on the mean parameter, and a zero for the second constraint, restricting the variance to be a positive value (>0). Otherwise, for small variances, the optimization routine might use nonpositive values of the variance to maximize the log-likelihood function. Attempting to take the log of zero (which is undefined) would cause the routine to terminate unsuccessfully.

Output – Example 8.5:

```
                Analysis variable : y


        Nobs           MEAN           VAR
        ------------------------------------

        100         4.23679        8.66171
        ------------------------------------
```

The above output results from the SUMMARY function, containing the sample mean and variance (LS estimates). By multiplying the variance by $(n - 1)/n = .99$ the LS variance estimate is converted to a value similar to that expected for the ML variance estimate, 8.5750929.

Output – Example 8.5 (continued):

```
            Newton-Raphson Optimization with Line Search
                     Without Parameter Scaling
           Gradient Computed by Finite Differences
        CRP Jacobian Computed by Finite Differences
              Parameter Estimates           2
              Lower Bounds                   1
              Upper Bounds                   0
                    Optimization Start
Active Constraints                    0  Objective Function        -272.5574242
Max Abs Gradient Element      14.91192627

                                       Objective   Max Abs           Slope of
              Function    Active       Function   Gradient   Step     Search
Iter Restarts  Calls   Constraints    Function     Change    Element  Size  Direction
  1      0       2           0        -255.23147   17.3260   11.7272  0.100  -235.9
  2*     0       3           0        -249.72501    5.5065    0.9299  1.000  -10.195
  3      0       4           0        -249.35015    0.3749    0.1366  1.000  -0.675
  4      0       5           0        -249.33698    0.0132    0.00576 1.000  -0.0256
  5      0       6           0        -249.33695   0.000024  0.000011 1.000  -488E-7
  6      0       8           0        -249.33695   9.77E-11  3.642E-7 1.003  -19E-11
                        Optimization Results
Iterations                   6   Function Calls                   9
Hessian Calls                7   Active Constraints               0
Objective Function  -249.336951   Max Abs Gradient Element    3.6422104E-7
Slope of Search Direction  -1.89776E-10  Ridge                  0
GCONV convergence criterion satisfied.
```

The above output results from the NLPNRA call. It contains information regarding the Newton-Raphson Optimization with Line Search. This routine is a numerical methodology used to find the gradient of the log-likelihood function with a slope of near zero, indicating a minimum, maximum, or saddle point in the function. The second derivative or a Hessian matrix is then used to determine whether the point is a minimum or a maximum.

The first portion of the output indicates that two parameter estimates were sought and that one lower bound was included as a constraint. That portion of the output can be used to verify that constraints were accurately included and that the correct number of parameter estimates was sought. Next, it can be seen that six iterations were employed to identify the maximum of the log-likelihood function. Maximization is successful when the change in the objective function is very small, from one iteration to the next, and the gradient and slope having values close to zero.

Output – Example 8.5 (continued):

```
       Reason for termination of optimization:   6


                  MEAN         VARIANCE


        ML Estimates:  4.2367888  8.5750938


      Maximum of Log-likelihood Function:   -249.337
```

The final portion of the output results from the print statement. The reason for termination of optimization has a reason code of 6. Positive reason codes indicate the termination of the optimization is due to a criterion being satisfied. Negative reason codes indicate unsuccessful termination.

8.4 Analysis of Variance

The analysis of variance (ANOVA) entails a comparison of systematic variability across treatment factors to the random variability among subjects within treatment groups. The test then determines whether systematic variability across treatments significantly exceeds that of the variability that might be caused by chance. The examples in this section are of the most basic fixed-factor ANOVA models. These examples can then be modified to meet the researcher's needs. Comparisons are also made to PROC GLM, which is the tool of choice for analyzing the fixed-factor data in practice. Although the methods employed in the PROC IML code are not identical to those in the PROC GLM code, the PROC IML examples are useful for academic purposes and the resulting numeric values obtained by both PROCs are very similar.

8.4.1 Basic One-Way ANOVA

Example 8.6 – One-way ANOVA:

A researcher is interested in studying the effect of three different teaching methods on the reading speed of students. The following data were collected:

	Teaching Method	
Method 1	Method 2	Method 3
298	298	309
306	310	317
289	303	313
295	301	300
291	299	304

Relevant formulas to one-way ANOVA include the following:

$I_n = n \times n$ identity matrix.
$J_n = n \times n$ matrix of 1's.

$X^- = \text{MP - inverse.}$

$$SSW = Y' \left(I_n - XX^- \right) Y$$

$$SST = Y' \left(I_n - \frac{1}{n} J_n \right) Y$$

$$SSB = Y' \left(XX^- - \frac{1}{n} J_n \right) Y = SST - SSW$$
$$d.f._{within} = n - p$$

$$d.f._{between} = p - 1$$

$$d.f._{total} = n - 1$$

$$MSB = \frac{SSB}{d.f._{between}}$$

$$MSW = \frac{SSW}{d.f._{within}}$$

$$F = \frac{MSB}{MSW}$$

P value $= P(f_{0.05,dfb,dfw} > F)$

The following SAS code can be used to conduct one-way ANOVA using Proc IML:

```
DATA reading;
    INPUT method score @@;
DATALINES;
1    298 1    306 1    289 1    295 1    291 2    298 2    310 2 303
2    301 2    299 3    309 3    317 3    313 3    300 3    304
;
RUN;
PROC IML;
    USE reading;
    READ ALL VAR {score} INTO score;
    READ ALL VAR {method} INTO method;
    y=score;
    n=NROW(y);
    x=J(n,1,1)||DESIGN(method); /**creation of design matrix**/
    xgx=x*GINV(x);
    rank=ROUND(TRACE(xgx));
    ssw=y'*(I(n)-xgx)*y; /** Sums of Squares computations **/
    ssb=y'*(xgx-(1/n)*J(n))*y;
    sst=y'*(I(n)-(1/n)*J(n))*y;
    ssb2=sst-ssw;
    dfw=n-rank; /** degrees of freedom computations **/
    dfb=rank-1;
    dft=n-1;
    msb=ssb/dfb; /** computations of mean squares **/
    msw=ssw/dfw;
    f=msb/msw; /** computations of f-statistic **/
    p_value=1-PROBF(f,dfB,dfw); /** computation of P value **/
    source={"Between","Within","Total"};
    ss=ssb//ssw//sst;
    df=dfb//dfw//dft;
```

```
    ms=msb//msw//mst;
    PRINT "ANOVA Table",
           source ss df ms f p_value,,,ssb2;
QUIT;
```

The between-treatment sums of squares have been computed using two different methods resulting in **SSB** and **SSB2**. The first, **SSB**, appears in the ANOVA Table in the output and the second, **SSB2**, appears below the ANOVA Table in the output. Notice that they both have exactly the same value. Some textbooks present both methods so that the reader can see that there are two methods for computing *SSB*. When computing sums of squares by hand in an academic setting, subtracting *SSW* from *SST* is far easier than computing *SSB* using a quadratic form directly. Here we see that both methods for computing *SSB* result in the same value.

Output – Example 8.6: ANOVA Table using IML

```
                    ANOVA Table

      source       ss        df       ms          f        p_value
      Between      409.6      2        204.8  5.4036939  0.0212146
      Within       454.8      12        37.9
      Total        864.4      14
                                                ssb2
                                                409.6
```

Example 8.7 – One-way ANOVA using GLM:

The following GLM code can be used instead of the IML procedure in the previous example to analyze the data from the previous example.

```
PROC GLM DATA=reading;
    CLASS method;
    MODEL score=method;
RUN;QUIT;
```

A portion of the output for the GLM procedure is given below. However, the GLM output contains other important statistics that are useful in data analysis.

Output – Example 8.7: ANOVA Table using GLM

```
                         The GLM Procedure
Dependent Variable: score
                                    Sum of
   Source            DF      Squares   Mean Square  F Value  Pr > F
   Model              2  409.6000000   204.8000000    5.40   0.0212
   Error             12  454.8000000    37.9000000
   Corrected Total   14  864.4000000
```

Notice that the numbers in the above output are the same as those for the output in the previous example.

8.4.2 Estimable Functions

An issue of great importance in estimation and testing with regard to linear models is that of estimability. Consider

$$y_{ij} = \mu + \alpha_1 + \alpha_2 + \alpha_3 + \varepsilon_{ij},$$

a special case of the model in Equation 8.1 This model is used in Example 8.6 with the teaching methods. In this case, y_{ij} represents the responses with i indexing the method (1, 2, or 3) and j indexing the subjects assigned to each method. The μ represents the unknown mean response of all experimental units. The α_1, α_2, and α_3, represent the unknown fixed effect of teaching methods 1, 2, and 3, respectively. The ε_{ij} represents the error variability with i indexing the method (1, 2, or 3) and j indexing the subjects assigned to each method. The unknown values μ, α_1, α_2, and α_3, are represented by the vector β in Equation 8.1. The design matrix, X, in Equation 8.1 consists of a column for each of the rows of the vector β, with the first column simply being a column of ones. Because that column of ones does not add to the rank of the matrix X, the rank of the matrix X in this example is equal to three, the number of teaching methods. The addition of the column of ones creates an issue with estimability. Not all linear functions of β are estimable, and consequently testable.

Graybill (2000) showed that a linear function $k'\beta$ is estimable if $rank[X', k] = rank[X']$, where $[X', k]$ is the side-by-side concatenation of X' and k. Consider the following values of k:

$$k1 = \begin{bmatrix} 1 \\ 0 \\ 0 \\ 0 \end{bmatrix}, \; k2 = \begin{bmatrix} 0 \\ 1 \\ 0 \\ 0 \end{bmatrix}, \; k3 = \begin{bmatrix} 1 \\ 1 \\ 0 \\ 0 \end{bmatrix}, \; k4 = \begin{bmatrix} 0 \\ 1 \\ -1 \\ 0 \end{bmatrix}$$

It will be shown in the following SAS example that $\beta'k1$ and $\beta'k2$ are not estimable functions; however, $\beta'k3$ and $\beta'k4$ are estimable functions.

Related to estimable functions are testable functions. Graybill (2000) showed that the hypothesis

$$H_0{:}k'\beta = 0 vs. H_0{:}k'\beta \neq 0$$

is testable if the linear function $k'\beta$ represents a linearly independent set of estimable functions. Again, one can determine if $rank[X', k] = rank[X']$, where k can, and often does, contain more than one column as long as all of its columns are linearly independent (full-column rank). Consider the following values of k:

$$k5 = \begin{bmatrix} 1 & 0 \\ 1 & 1 \\ 0 & 0 \\ 0 & 0 \end{bmatrix}, \ k6 = \begin{bmatrix} 1 & 1 & 1 \\ 1 & 0 & 0 \\ 0 & 1 & 0 \\ 0 & 0 & 1 \end{bmatrix}$$

It will be shown in the following SAS example that $k5$ is not testable because the second column (which is the same as $k2$) does not produce an estimable function. Consider the following example:

Example 8.8 – Estimable (and testable) functions:

The following code

```
DATA reading;
    INPUT method score @@;
DATALINES;
1    298 1    306 1    289 1    295 1    291 2    298 2    310 2 303
2    301 2    299 3    309 3    317 3    313 3    300 3    304
;
RUN;
PROC IML;
    USE reading;
    READ ALL VAR {score} INTO score;
    READ ALL VAR {method} INTO method;
    y=score;
    n=NROW(y);
    x=J(n,1,1)||DESIGN(method);
    /** creation of design matrix **/
    START matrank(x); /** matrank computes matrix rank **/
        matrank=ROUND(TRACE(x*GINV(x)));
        RETURN (matrank);
    FINISH;
    rankx=matrank(x);
    /** Define linear function coefficient
      vectors and matrices **/
    k1={1,0,0,0};
    k2={0,1,0,0};
    k3={1,1,0,0};
    k4={0,1,-1,0};
    k5={1 1,0 1,0 0,0 0};
```

```
   k6={1 1 1,1 0 0,0 1 0,0 0 1};
   /** check for estimability **/
   START est(x,k);
      xk=x'||k;
      IF matrank(xk)=matrank(x) & matrank(k)=NCOL(k)
      THEN est="estimable";
      ELSE est="not estimable";
      RETURN (est);
   FINISH;
   PRINT "k1 is" (est(x,k1));
   PRINT "k2 is" (est(x,k2));
   PRINT "k3 is" (est(x,k3));
   PRINT "k4 is" (est(x,k4));
   PRINT "k5 is" (est(x,k5));
   PRINT "k6 is" (est(x,k6));
QUIT;
```

The EST modular function tests for estimability by checking to see if the rank of the concatenated matrix **XK** is the same as that of the matrix **X**, and by checking to see that if there are multiple columns in the linear function that those columns are linearly independent. This modular function, or a variation, appears on all the subsequent examples when there is cause to question the estimability/testability of a linear function. The MATRANK function is created to compute a matrix rank.

Output – Example 8.8:

```
              k1 is not estimable
              k2 is not estimable
              k3 is estimable
              k4 is estimable
              k5 is not estimable
              k6 is estimable
```

Note that in the output, the word "estimable" also implies "testable."

The SAS procedures GLM, MIXED, etc. check for estimability and testability and will provide messages in the output to inform the user when non-estimable (or non-testable) functions have been included in the analysis and will not provide estimates or test results for such functions. If producing linear functions and tests in the IML procedure, non-estimable functions can be used to compute results; however, such results will not be of value in a proper analysis. Checking for estimability can help avoid this situation.

8.4.3 Two-Way ANOVA with Interaction

Example 8.9 – Two-way ANOVA using IML:

A researcher is interested in studying the effect of gender, three different teaching methods, and their possible interaction on the vocabulary test scores of the students. The following data were collected:

Method	Gender	Score
1	1	49
1	1	54
1	1	43
1	2	47
1	2	44
1	2	44
2	1	50
2	1	58
2	1	53
2	2	52
2	2	51
2	2	52
3	1	56
3	1	60
3	1	58
3	2	50
3	2	53
3	2	54

Relevant formulas to two-way ANOVA include the following:

$I_n = n \times n$ identity matrix.

$J_n = n \times n$ matrix of 1's.

$X^- = $ MP-inverse.

$X = [X_A | X_B | X_{AB}]$ is a partitioned matrix such that X_A is $n \times a$, X_B is $n \times b$, and X_{AB} is $n \times a*b$; where a is the number of levels of *method* (Factor A) and b is the number of levels of *gender* (Factor B).

$$SSW = Y' \left(I_n - XX^-\right) Y$$

$$SST = Y' \left(I_n - \frac{1}{n} J_n\right) Y$$

$$SSBetween = SST - SSW$$

$$SSA = Y' \left(X_A X_A^- - \frac{1}{n} J_n\right) Y$$

$$SSB = Y' \left(X_B X_B^- - \frac{1}{n} J_n \right) Y$$

$$SSAB = SST - SSA - SSB - SSW$$

$$d.f._{within} = n - p$$

$$d.f._{between} = p - 1$$

$$d.f._{total} = n - 1$$

$$d.f._A = a - 1$$

$$d.f._B = b - 1$$

$$d.f._{AB} = (a - 1)(b - 1)$$

$$MSBetween = \frac{SSBetween}{d.f._{between}}$$

$$MSW = \frac{SSW}{d.f._{within}}$$

$$MSA = \frac{SSA}{d.f._A}$$

$$MSB = \frac{SSB}{d.f._B}$$

$$MSAB = \frac{SSAB}{d.f._{AB}}$$

$$F_{overall} = \frac{MSBetween}{MSW}$$

$$F_A = \frac{MSA}{MSW}$$

$$F_B = \frac{MSB}{MSW}$$

$$F_{AB} = \frac{MSAB}{MSW}$$

$$Pvalue(\text{overall}) = P(f_{0.05, \text{dfbetween, dfw}} > F)$$

$$Pvalue(A) = P(f_{0.05, \text{dfA, dfw}} > F)$$

$$Pvalue(B) = P(f_{0.05, \text{dfB, dfw}} > F)$$

$$Pvalue(AB) = P(f_{0.05, \text{dfAB, dfw}} > F)$$

The following SAS code can be used to conduct two-way ANOVA (balanced case) using Proc IML. This code demonstrates the automation of the linear models formulas associated a two-way ANOVA. It also provides another example of the use of the DESIGN function.

```
DATA reading;
    INPUT a b score;
    c=a*b;
    DATALINES;
1    1    49
1    1    54
1    1    43
1    2    47
1    2    44
1    2    44
2    1    50
2    1    58
2    1    53
2    2    52
2    2    51
2    2    52
3    1    56
3    1    60
3    1    58
3    2    50
3    2    53
3    2    54
;
RUN;

PROC IML;
    USE reading;
    READ ALL VAR {score} INTO score;
    READ ALL VAR {a} INTO a;
    READ ALL VAR {b} INTO b;
    READ ALL VAR {c} INTO c;
    y=score;
    xa=DESIGN(a);
    xb=DESIGN(b);
    xc=DESIGN(c);
    x=xa||xb||xc;
    n=NROW(x);
    START matrank(x);   /** matrank computes matrix rank **/
        matrank=ROUND(TRACE(x*GINV(x)));
```

```
      RETURN (matrank);
FINISH;
rankx=matrank(x);
ranka=matrank(xa);
rankb=matrank(xb);
ssa=y'*(xa*GINV(xa)-(1/n)*J(n))*y;
/** compute sums of squares **/
ssb=y'*(xb*GINV(xb)-(1/n)*J(n))*y;
sse=y'*(I(n)-x*GINV(x))*y;
sst=y'*(I(n)-(1/n)*J(n))*y;
ssab=sst-ssa-ssb-sse;
ssm=sst-sse;
dfa=ranka-1; /** compute degrees of freedom **/
dfb=rankb-1;
dfab=dfa*dfb;
dfe=n-rankx;
dft=n-1;
dfm=rankx-1;
msa=ssa/dfa; /** compute mean squares **/
msb=ssb/dfb;
msab=ssab/dfab;
msm=ssm/dfm;
mse=sse/dfe;
f=msm/mse; /** compute f-statistics **/
f_a=msa/mse;
f_b=msb/mse;
f_ab=msab/mse;
p_value=1- PROBF(f,dfm,dfe); /** compute P values **/
p_value_a=1-PROBF(f_a,dfa,dfe);
p_value_b=1-PROBF(f_b,dfb,dfe);
p_value_ab=1-PROBF(f_ab,dfab,dfe);
source={"Model","Error","Total"}; /** organize output **/
ss=ssm//sse//sst;
df=dfm//dfe//dft;
ms=msm//mse//mst;
PRINT "ANOVA Table",
      source ss df ms f p_value;
source={"A","B","AB"};
ss=ssa//ssb//ssab;
df=dfa//dfb//dfab;
ms=msa//msb//msab;
f=f_a//f_b//f_ab;
p_value=p_value_a//p_value_b//p_value_ab;
PRINT "ANOVA Table",
```

```
    source ss df ms f p_value;
QUIT;
```

Output – Example 8.9: Two-Way ANOVA Table using IML

ANOVA Table

source	ss	df	ms	f	p _value
Model	293.77778	5	58.755556	6.0434286	0.0050912
Error	116.66667	12	9.7222222		
Total	410.44444	17			

ANOVA Table

source	ss	df	ms	f	p _value
A	219.44444	2	109.72222	11.285714	0.001749
B	64.222222	1	64.222222	6.6057143	0.0245384
AB	10.111111	2	5.0555556	0.52	0.6073261

Example 8.10 Two-way ANOVA using GLM:

The following GLM code can be used instead of the IML procedure in the previous example to analyze the data from the previous example.

```
PROC GLM DATA=reading;
   CLASS a b;
   MODEL score=a|b;
RUN;QUIT;
```

A portion of the output for the GLM procedure is given below. However, the GLM output contains other important statistics that are useful in data analysis.

Output – Example 8.10:

The GLM Procedure

Dependent Variable: score

Source	DF	Sum of Squares	Mean Square	F Value	Pr > F
Model	5	293.7777778	58.7555556	6.04	0.0051
Error	12	116.6666667	9.7222222		
Corrected Total	17	410.4444444			

	R-Square	Coeff Var	Root MSE	score Mean	
	0.715755	6.047938	3.118048	51.55556	

Source	DF	Type I SS	Mean Square	F Value	Pr > F
A	2	219.4444444	109.7222222	11.29	0.0017
B	1	64.2222222	64.2222222	6.61	0.0245
A*B	2	10.1111111	5.0555556	0.52	0.6073

Source	DF	Type III SS	Mean Square	F Value	Pr > F
A	2	219.4444444	109.7222222	11.29	0.0017
B	1	64.2222222	64.2222222	6.61	0.0245
A*B	2	10.1111111	5.0555556	0.52	0.6073

Notice that the numbers in the above output are the same as those for the output in the previous example. The analyses are similar, but use different methods. The GLM procedure incorporates efficiencies and ensures better numerical accuracy. Using the IML code in Example 8.9 can take more time to execute and occasionally will result in numerical values that are not accurate due to errors in finite precision arithmetic.

8.4.4 A Specific Case of the General Linear Model (Weighted Least Squares)

Example 8.11 – Weighted least squares:

In this example, the assumption on the error variance is relaxed a bit. The assumption from the previous example,

$$\text{var}(\varepsilon) = \sigma^2 I_n,$$

will be relaxed to

$$\text{var}(\varepsilon) = \sigma^2 V,$$

where V is a known matrix of constants. This assumption is more flexible and allows for covariances that are scalar multiples of σ^2. This example demonstrates IML code that can be used for the weighted least squares (WLS) analysis of a specific case of general linear model

$$y = \beta_1 x_1 + \beta_2 x_2 + \beta_3 x_3 + \varepsilon,$$

where ε is normally distributed with mean 0 and variance $\sigma^2 V$. Specifically, the V matrix contains scalar values called "weights" for its diagonal elements and zeros for its off-diagonal elements. It is of interest to estimate

$$\beta = \begin{bmatrix} \beta_1 \\ \beta_2 \\ \beta_3 \end{bmatrix},$$

a 3×1 column vector of unknown parameters, and σ^2. It is also of interest to perform tests of hypothesis and construct confidence intervals.

Relevant formulas to the general linear model include the following:

$$\hat{\beta} = (X'V^{-1}X)^{-1}X'V^{-1}Y$$

$$\hat{\sigma}^2 = \frac{1}{n - Rank(X)} Y'[V^{-1} - V^{-1}X(X'V^{-1}X)^{-1}X'V^{-1}]Y$$

$$k'\hat{\beta}$$

$$SE(k'\hat{\beta}) = \hat{\sigma}\sqrt{k'(X'V^{-1}X)^{-1}k}$$

$$t = \frac{k'\hat{\beta} - k'\beta}{SE(k'\hat{\beta})}$$

$$F = \frac{(H\hat{\beta} - h)'[H(X'V^{-1}X)H']^{-1}(H\hat{\beta} - h)}{\hat{\sigma}^2(Rank(H))}$$

```
PROC IML;
   x={-1 -1 -1,-1 -1 1,1 -1 1,-1 1 -1,1 1 -1,1 1 1,
      1 -1 -1,-1 1 1};
   y={2,1,3,3,2,3,1,4};
   v=DIAG({1,.5,.25,.125,.0625,.03125,.015625,.0078125});
   START matrank(x); /** matrank computes matrix rank **/
      matrank=ROUND(TRACE(x*GINV(x)));
      RETURN (matrank);
   FINISH;
   n=NROW(x);
   p=NCOL(x);
   b=INV(x'*INV(v)*x)*x'*INV(v)*y;
   s2=(1/(n-p))*y'*(INV(v)-INV(v)*x*INV(x'*INV(v)*x)
      *x'*INV(v))*y;
   cov_B=s2*INV(x'*INV(v)*x);
   k={1,0,-3};
   /** Create Module: est to check for estimability **/
   START est(x,k);
      xk=x'||k;
      IF matrank(xk)^= matrank(x) | matrank(k)^= NCOL(k)
      THEN PRINT "Warning: Function" k "is not estimable!";
   FINISH;
   /** check for estimability of k **/
   CALL est(x,k);
   kpb=k'*b;
   se_kpb=SQRT(k'*cov_b*k);
   t_kpb=kpb/se_kpb;
   abs_t=ABS(t_kpb);
   p_val1=2*(1-PROBT(abs_t,n-p));
   ll=kpb-se_kpb*TINV(.975,n-p);
```

```
ul=kpb+se_kpb*TINV(.975,n-p);
h={0 1 0,0 0 1};
/** check for estimability of h **/
CALL est(x,h');
lilh={0,0};
q=matrank(h);
f=(h*b-lilh)'*INV(h*INV(x'*INV(v)*x)*h')*(h*b-lilh)/(q*s2);
p_val2=1-PROBF(f,q,n-p);
PRINT n p b s2,k kpb SE_kpb t_kpb p_val1 ll ul,
    h lilh q f p_val2;
QUIT;
```

Output – Example 8.11:

n	p	b	s2
8	3	0.2730015	240.60814
		1.9736048	
		0.6282051	

The first section of output includes the sample size, **N**; the number of unknown parameters in the column vector β, **P**; parameter estimates for β, **B**; and the estimate of d variance of σ^2, **S2**.

Output – Example 8.11 (continued):

k	kpb	se_kpb	t_kpb	p_val1	ll	ul
1	-1.611614	4.5466646	-0.354461	0.7374531	-13.29919	10.07596
0						
-3						

The second section of output includes a coefficient vector, **K**, which, when combined with the **B** vector, yields the estimate of $k'\beta = \beta_1 - 3\beta_3$, identified as **KPB**. Also included is the estimated standard error of that estimate, the t-statistic and P value for testing the hypothesis that $\beta_1 = 3\beta_3$, and the lower and upper 95% confidence limits on $\beta_1 = 3\beta_3$.

Output – Example 8.11 (continued):

h		lilh	q	f	p_val2
0	1	0	0	2	1.9372367	0.2382695
0	0	1	0			

The final section of output includes the **H** matrix and **LILH** column vector for testing the hypothesis

$$H_0: \beta_2 = \beta_3 = 0.$$

Also included are the rank of the **H** matrix, **Q**; and the F-statistic and P value for the test.

This model is considered "full-rank" model because the $X'V^{-1}X$ matrix is a $p \times p$ matrix of rank p. A similar analysis could have been run with a "less-than-full rank" model. In that case the $X'V^{-1}X$ matrix would have been a $p \times p$ matrix of rank $v < p$. To work with models that are less-than-full-rank, simply change all of the INV functions to GINV functions. This would effectively use the MP-inverse to compute estimates instead of inverses. Note that though the code for this example is used for weighted least squares, it would also work for a more general V matrix of known constants.

Example 8.12 – ANOVA using GLM:

The following analysis uses the GLM procedure to analyze the data given in the previous example using weighted least squares:

```
DATA data1;
   w2=1/w;
   INPUT y x1 x2 x3 w;
   DATALINES;
   2 -1 -1 -1 1
   1 -1 -1 1 .5
   3 1 -1 1 .25
   3 -1 1 -1 .125
   2 1 1 -1 .0625
   3 1 1 1 .03125
   1 1 -1 -1 .015625
   4 -1 1 1 .0078125
   ;
RUN;

PROC GLM DATA=data1;
   MODEL y=x1 x2 x3/SOLUTION NOINT;
   WEIGHT w2;
   ESTIMATE 'kpb' x1 1 x2 0 x3 -3;
   CONTRAST 'x2=x3' x2 1,x3 -1;
QUIT;
```

Output – Example 8.12:

Source	DF	Sum of Squares	Mean Square	F Value	Pr > F
Model	3	1374.959276	458.319759	1.90	0.2468
Error	5	1203.040724	240.608145		
Uncorrected Total	8	2578.000000			

R-Square	Coeff Var	Root MSE	y Mean
0.533343	531.6458	15.51155	2.917647

NOTE: No intercept term is used: R-square is not corrected for the mean.

Source	DF	Type I SS	Mean Square	F Value	Pr > F
x1	1	442.7294118	442.7294118	1.84	0.2330
x2	1	896.0156863	896.0156863	3.72	0.1115
x3	1	36.2141780	36.2141780	0.15	0.7140

Source	DF	Type III SS	Mean Square	F Value	Pr > F
x1	1	10.5414982	10.5414982	0.04	0.8425
x2	1	393.5180062	393.5180062	1.64	0.2571
x3	1	36.2141780	36.2141780	0.15	0.7140

Contrast	DF	Contrast SS	Mean Square	F Value	Pr > F
x2=x3	2	932.2298643	466.1149321	1.94	0.2383

Parameter	Estimate	Standard Error	t Value	Pr > |t|
kpb	-1.61161388	4.54666461	-0.35	0.7375

Notice that the numbers in the above output are the same as those for the output in the previous example. As previously mentioned, the GLM procedure (Example 8.10) implements numerical efficiencies that are missing from the IML code (Example 8.11).

8.5 Multiple Linear Regression Analysis

Linear regression is a special case of a linear model. The linear regression equation has the following form:

$$y = \beta_0 + \beta_1 x_1 + \beta_2 x_2 + \cdots + \beta_k x_k + \varepsilon, \qquad (8.5)$$

where y represents the response variable, $\beta_0, \beta_1, \ldots, \beta_k$ represent unknown parameters, $x_1, x_2, \cdots, x_k$ represent the regressor variables, and ϵ represents the random error. The assumptions with the model that allow for LS estimation and hypothesis testing are that the model has been appropriately and completely specified and that the errors are normally and independently distributed with mean equal to zero and a constant variance.

Equation 8.5 can be written as a linear model (Equation 8.1) such that **Y** is a $n \times 1$ column vector of response values, a $n \times p$ matrix **X** constructed by concatenating a $n \times 1$ column of ones to the left of the $n \times k$ matrix of regressor values ($p = k + 1$), β is the $p \times 1$ column vector of unknown parameter values, and ε is the $n \times 1$ column vector of unknown random errors.

As with analysis of variance, multiple regression analysis involves partitioning the response data variability to determine if a significant portion of the variability associated with the response (dependent) variable can be attributed to a phenomenon under study. With ANOVA for a designed experiment, the phenomenon under study may be an applied treatment. With regression, the phenomenon under study may be associated with a regressor (independent) variable or a function of multiple regressor variables. Consequently, the total variability associated with the response variable may be partitioned a number of different ways to consider a number of different phenomena. One way to partition the variability is to separate the variability due to the model as defined from the remainder of the variability, referred to as the residual.

- **Regression sum of squares** (SSR) represents the amount of variability in the data due to the model.

$$SSR = Y'\left(XX^- - \frac{1}{n}J_n\right)Y$$

The degrees of freedom associated with SSR is the rank of

$$\left(XX^- - \frac{1}{n}J_n\right), \ d.f._R = rank(X) - 1.$$

- **Error (or "residual") sum of squares** (SSE) represents the amount of variability in the data remaining after the model is taken into consideration.

$$SSE = Y'(I_n - XX^-)Y$$

The degrees of freedom associated with SSE is the rank of $(I_n - XX^-)$, $d.f._E = n - rank(X)$.
- **Total (corrected) sum of squares** (SST) represents the total amount of variability in the data, comprising variability due to the model (SSR) and remaining variability (SSE).

$$SST = Y'\left(I_n - \frac{1}{n}J_n\right)Y$$

The degrees of freedom associated with SST is the rank of $(I_n - \frac{1}{n}J_n)$, $d.f._\tau = n - 1$.

A function of the sums of squares can be used to test the model significance. A ratio of the sums of squares, divided by their respective degrees of freedom constitutes an F distribution.

$$F = \frac{SSR/d.f._R}{SSE/d.f._R} \tag{8.6}$$

Using sample data in Equation 8.6 creates a test statistic for the test of model significance. The test statistic is compared to a value from the F distribution with degrees of freedom $d.f._R$ and $d.f._E$.

A linear function of regressor parameters can also be tested. The linear function is of the type $L'\beta = L_0\beta_0 + L_1\beta_1 + \cdots + L_k\beta_k$, made up of the $p \times 1$ column vector L and the $p \times 1$ column vector β. The test statistic for testing the hypothesis

$$H_0 : L'\beta = 0$$

is defined as

$$t = \frac{L'\hat{\beta}}{\sqrt{MSE \times L'(X'X)^- L}},$$

where

$$MSE = \frac{SSE}{d.f._E}.$$

Multiple linear functions can be tested simultaneously. For this situation, L is defined as a $m \times 1$ column vector of m $1 \times p$ row vectors or linear functions, consequently forming a $m \times p$ matrix to be renamed H to avoid confusion. The test statistic for testing the hypothesis

$$H_0 : H\beta = 0$$

is defined as a ratio with numerator

$$numerator = \frac{(H\hat{\beta})'[H(X'X)^{-1}H']^{-1}(H\hat{\beta})}{rank(H)}$$

and MSE as the denominator.

The test statistic is computed as

$$F = \frac{numerator}{denominator}$$

and is compared to a value from the F distribution with $rank(H)$ and $d.f._E$ degrees of freedom.

The variance inflation factor (VIF) for a given regressor, i, is computed as

$$VIF_i = \frac{1}{1 - R_i^2} = \frac{SST_i}{SSE_i}$$

where R_i^2 represents the coefficient of multiple determination obtained by regressing X_i on all other regressor variables. Likewise, SST_i and SSE_i respectively represent the total sum of squares and error sum of squares obtained by regressing X_i on all other regressor variables. The VIF statistic is a multicollinearity diagnostic that measures the increase in variances of the regressor estimates caused by the linear relationship among the regressor variables. A VIF value of 10, for example, represents a ten-fold increase in the variances of the estimates due to the multicollinearity. Inflated variances can lead to issues like a reduction in the sensitivity of hypothesis tests related to the regressor coefficients.

Example 8.13 – A linear regression model:

The following data represent measurements taken on a group of students enrolled in a course on introductory statistics. It is desirable to know how well height, weight, and shoe size can be used to estimate final exam scores—the assumption is they can't. The chosen model to estimate is

$$y = \beta_0 + \beta_1 x_1 + \beta_2 x_2 + \beta_3 x_3 + \varepsilon,$$

represented by **Y=B0+B1X1+B2X2+B3X3+E**. **Y** is a 40 × 1 vector of recorded final exam scores. The heights are recorded in **X1**, the weights in **X2**, shoe sizes in **X3**, and **E** represents the random model error. So, **X** is a 40 × 3 matrix, and **E** is a 40 × 1 vector with mean of **0** and variance $\sigma^2 \mathbf{I}$. **B**={**B0,B1,B2,B3**}, a 4 × 1 column vector of unknown parameters. Because of the large amount of code for this example, the IML code is divided into smaller sections, as is the output. Each section will be preceded by an explanation.

In the first section of code, the PROC IML step begins with the assignment of the data to the **X** matrix and the **Y** column vector. The statistics that make up the ANOVA table are created and then printed. Formatting is used in the PRINT statement to make the output look more similar to the PROC REG output in the following example. The creation of some of the variables is to save repetitive steps. For example, the **XPXI** matrix involves the computation of a MP-inverse and is used in several places throughout the code. The result is used throughout the code, though the computation occurs only once. Likewise, degrees of freedom are assigned to a variable and then are used throughout the code.

```
PROC IML;
  x={71 183 11.0,69 125 9.0,71 145 11.0,69 125 8.5,74 180 9.0,
     67 135 7.0, 76 185 15.0,68 120 8.5,66 112 7.5,66 137 9.5,
     64 110 8.0,66 150 8.5,64 100 7.5,71 196 10.5,65 120 8.5,
     67 125 9.0,66 165 9.5,67 145 8.5,72 170 11.0,70 185 8.5,
     69 190 9.0,69 186 12.0,69 145 10.0,68 150 9.5,66 140 8.5,
     67 175 10.5,68 152 9.5,67 155 8.5,69 150 9.5,73 175 11.0,
     70 170 13.0,74 192 12.0,71 165 10.5,61 100 7.5,
     68 160 10.0,72 175 12,62 110 6.5,64 110 7.0,68 140 10.0};
  y={136,179,188,140,164,165,178,154,186,169,179,153,174,186,
     177,179,189,196,149,181,163,130,164,192,170,177,179,176,
     180,183,164,191,172,133,165,133,179,183,198};
  n=NROW(x);
  x=J(n,1)||x;
```

```
xpxi=GINV(x`*x);
START matrank(x); /** matrank computes matrix rank **/
    matrank=ROUND(TRACE(x*GINV(x)));
    RETURN (matrank);
FINISH;
rank_x=matrank(x);
dfr=rank_x-1;
dfe=n-rank_x;
dft=n-1;
b_hat=GINV(x)*y;
p=NCOL(x);
resid=y - x*b_hat;
y_hat=x*b_hat;
sstot=y`*y - y`*((1/n)*J(n))*y;
sse=y`*y - b_hat`*x`*y;
ssr=sstot-sse;
dfe=n-rank_x;
msr=ssr/dfr;
mse=sse/dfe;
r2=(SSQ(y-SUM(y)/n)-SSQ(resid))/(SSQ(y-SUM(y)/n));
f=msr/mse;
p_val_f=1-PROBF(f,dfr,dfe);
/** The following vectors are created for output
    purposes **/
source={"Model","Error","Corrected Total"};
df=dfr//dfe//dft;
ss=ssr//sse//sstot;
ms=msr//mse;
PRINT y y_hat [FORMAT=8.2] resid [FORMAT=8.2],
      "Analysis of Variance",source df ss [FORMAT=8.4]
         ms [FORMAT=8.4]
      f [FORMAT=7.2] p_val_f [FORMAT=8.4],r2 [FORMAT=6.4];
```

The second section of code computes the parameter estimates and their associated standard errors, test statistics, and P values for their respective tests of significance. Additionally, the functional module VIF is created to compute the variance inflation factor (VIF) for a given regressor. The column vector **X1** contains the data for the regressor variable of interest. The matrix **X2** contains the data for the remaining regressor variables (not in **X1**). The statement

```
x2=x[,LOC(1:p^=i+1)];
```

uses the LOC function to identify all the column numbers in the **X** matrix other than the one used in the column vector **X1**. The total sum of squares, **SSTOT2**, and error sum of squares, **SSE2**, are computed from regressing **X1** on **X2**. The ratio **SSTOT2/SSE2** results in the VIF of interest.

```
var_b=VECDIAG(mse*xpxi);
   se_b=SQRT(var_b);
   t_b=b_hat/se_b;
   p_val_b=2*(1-PROBT(ABS(t_b),dfe));
   START vif(x,i);
       n=NROW(x);
       p=NCOL(x);
   x1=x[,i+1];
   x2=x[,LOC(1:p^=i+1)];
   b_hat=INV(x2'*x2)*x2'*x1;
       sstot2=x1'*x1 - x1'*((1/n)*J(n))*x1;
       sse2=x1'*x1 - b_hat'*x2'*x1;
       vif=sstot2/sse2;
         RETURN(vif);
   FINISH;
   vif0=vif(x,0);
   vif1=vif(x,1);
   vif2=vif(x,2);
   vif3=vif(x,3);
   vif=vif0//vif1//vif2//vif3;
   variable={"Intercept",x1,x2,x3};
   PRINT "Parameter Estimates",variable b_hat [FORMAT=8.4]
           se_b [FORMAT=8.4] t_b [FORMAT=7.2]
               p_val_b [FORMAT=8.4]
           vif [format=10.5];
```

The next section of code estimates and tests the significance of linear functions of the regressor parameters. The column vector **L** contains the coefficients for a single linear function. The linear function represents $\beta_2 = \beta_3$. The matrix **H** contains two rows of coefficients for a pair of linear functions. This pair of linear functions simultaneously represents $\beta_2 = 0$ and $\beta_3 = 0$. The format of the output for the second test more closely matches the output of PROC REG. The format of the output for the first test appears similar to that of the ESTIMATE statement in PROC GLM. Also, the first test uses the T distribution for testing where the second test uses the F distribution. The EST modular function is defined to test the linear function, **L**, for estimability. It will create a warning message if the function is not estimable. It will not create a message if the function is estimable.

```
   /** Linear function LB **/
   l={0,0,1,-1};
   /** Create Module: est to check for estimability **/
   START est(x,k);
     xk=x'||k;
       IF matrank(xk)^= matrank(x) | matrank(k)^= NCOL(k)
         THEN PRINT "Warning: function" k "is not estimable!";
   FINISH;
```

```
/** check for estimability of l **/
CALL est(x,l);

lb=l'*b_hat;
se_lb=SQRT(mse*l'*xpxi*l);
t_lb=lb/se_lb;
p_val_lb=2*(1-PROBT(ABS(t_lb),dfe));
estimate=lb;

/** Pair of Linear functions HB **/
h={0 0 1 0,
   0 0 0 1};

/** checking for estimability of h **/
CALL est(x,h');

rank_h=matrank(h);
hb=h*b_hat;
numerator=(hb'*GINV(h*xpxi*h')*hb)/rank_h;
denominator=mse;
f_hb=numerator/denominator;
p_val_hb=(1-PROBF(f_hb,rank_h,dfe));
source={"Numerator","Denominator"};
mean_squares=numerator//denominator;
PRINT "Test: B2=B3" estimate [FORMAT=8.4] se_lb
[FORMAT=8.4]
        t_lb [FORMAT=7.2] p_val_lb [FORMAT=8.4]
"Test: B2=0,B3=0" source mean_squares f_hb [FORMAT
=7.2]
        p_val_hb [FORMAT=8.4];
QUIT;
```

Output – Example 8.13:

y	y_hat	resid
136	168.83	-32.83
179	173.57	5.43
188	170.49	17.51
140	174.64	-34.64
164	176.32	-12.32
165	175.35	-10.35
178	165.35	12.65
154	173.82	-19.82
186	174.25	11.75
169	168.89	0.11
179	171.21	7.79
153	170.46	-17.46

174	172.71	1.29
186	169.34	16.66
177	170.74	6.26
179	171.51	7.49
189	167.67	21.33
196	171.71	24.29
149	170.43	-21.43
181	173.06	7.94
163	170.74	-7.74
130	164.51	-34.51
164	170.56	-6.56
192	170.38	21.62
170	170.90	-0.90
177	166.13	10.87
179	170.30	8.70
176	171.27	4.73
180	171.41	8.59
183	171.24	11.76
164	164.10	-0.10
191	169.39	21.61
172	170.68	1.32
133	169.62	-36.62
165	168.88	-3.88
133	168.07	-35.07
179	172.35	6.65
183	173.35	9.65
198	169.75	28.25

The above section of output includes the data values, **Y**; the predicted values, **Y_HAT**; and the residuals, **RESID**.

Output – Example 8.13 (continued):

Analysis of Variance

source	df	ss	ms	f	p_val_f
Model	3	285.0675	95.0225	0.276	0.842
Error	35	12052.16	344.3475		
Corrected Total	38	12337.23			

$$r2$$
$$0.0231$$

The above section of output contains the abbreviated ANOVA table for the model as well as the coefficient of determination, **R2**.

Output – Example 8.13 (continued):

```
                    Parameter Estimates
  variable      b_hat       se_b      t_b   p_val_b         vif

  Intercept   127.2010    90.6990    1.40    0.1696    -0.00000
  X1            1.0293     1.6977    0.61    0.5482     3.40075
  X2           -0.0435     0.1779   -0.24    0.8084     2.76240
  X3           -2.1359     2.8321   -0.75    0.4558     2.76462
```

The above section of output includes the parameter estimates, standard errors for these estimates, test statistics for comparing the parameter estimates to zero, and the associated *P* values.

Output – Example 8.13 (continued):

```
                       Test: B2=B3

        estimate     se_lb    t_lb p_val_lb

         2.0924     2.8876    0.72   0.4735
                 Test: B2=0,B3=0

     source        mean_squares     f_hb p_val_hb

     Numerator       137.27277      0.40   0.6742
     Denominator     344.34752
```

The above section of output entitled "Test: B2=B3" tests the hypothesis that **B2=B3** using a linear function $L'\beta$. The above section of output entitled "Test: B2=0, B3=0" simultaneously tests the hypotheses **B2=0** and **B3=0**.

Example 8.14 – Regression analysis using REG:

The following analysis uses the REG procedure to analyze the data given in the previous example. First, the matrix **X** and the column vector **Y** are created in the PROC IML step. Then they are concatenated together and assigned to the variable IMLREGDATA. The CREATE and APPEND statements combine to create the SAS data set REGDATA from the IML matrix IMLREGDATA. The model is then analyzed in the PROC REG step. The model is defined in the MODEL statement. The first TEST statement tests the hypothesis $\beta_2 = \beta_3$. The second test statement simultaneously tests the hypotheses $\beta_2 = 0$ and $\beta_3 = 0$.

```
PROC IML;
   x={71 183 11.0,69 125 9.0,71 145 11.0,69 125 8.5,74
      180 9.0,67 135 7.0,76 185 15.0,68 120 8.5,66 112 7.5,
      66 137 9.5,64 110 8.0,66 150 8.5,64 100 7.5,71 196 10.5,
      65 120 8.5,67 125 9.0,66 165 9.5,67 145 8.5,72 170 11.0,
      70 185 8.5,69 190 9.0,69 186 12.0,69 145 10.0,68 150 9.5,
      66 140 8.5,67 175 10.5,68 152 9.5,67 155 8.5,69 150 9.5,
```

```
   73 175 11.0,70 170 13.0,74 192 12.0,71 165 10.5,
   61 100 7.5,68 160 10.0,72 175 12,62 110 6.5,64 110 7.0,
   68 140 10.0};
y={136,179,188,140,164,165,178,154,186,169,179,153,174,186,
   177,179,189,196,149,181,163,130,164,192,170,177,179,176,
   180,183,164,191,172,133,165,133,179,183,198};
imlregdata=y||x;
CREATE regdata FROM imlregdata [COLNAME={y,x1,x2,x3}];
APPEND FROM imlregdata;
QUIT;

PROC REG DATA=regdata;
   MODEL y=x1 x2 x3/vif;
   TEST x2=x3;
   TEST x2=0,x3=0;
QUIT;
```

Output – Example 8.14:

The REG Procedure

Model: MODEL1

Dependent Variable: Y

Number of Observations Read	39
Number of Observations Used	39

Analysis of Variance

Source	DF	Sum of Squares	Mean Square	F Value	Pr > F
Model	3	285.06748	95.02249	0.28	0.8423
Error	35	12052	344.34752		
Corrected Total	38	12337			

Root MSE	18.55660	R-Square	0.0231	
Dependent Mean	170.61538	Adj R-Sq	-0.0606	
Coeff Var	10.87628			

Parameter Estimates

Variable	DF	Parameter Estimate	Standard Error	t Value	Pr > |t|	Variance Inflation
Intercept	1	127.20097	90.69896	1.40	0.1696	0
X1	1	1.02934	1.69769	0.61	0.5482	3.40075
X2	1	-0.04347	0.17790	-0.24	0.8084	2.76240
X3	1	-2.13588	2.83207	-0.75	0.4558	2.76462

```
                      Test 1 Results for Dependent Variable Y

                                        Mean
        Source               DF        Square      F Value     Pr > F

        Numerator             1       180.80714       0.53      0.4735
        Denominator          35       344.34752
                      Test 2 Results for Dependent Variable Y

                                        Mean
        Source               DF        Square      F Value     Pr > F

        Numerator             2       137.27277       0.40      0.6742
        Denominator          35       344.34752
```

By observation, it can be seen that the above results closely resemble those of the previous example. However, the IML code does not match the efficiency and numerical accuracy of the REG code.

8.6 Correlation Analysis

There are several different types of correlation of interest in linear model analysis. This chapter will examine four such types:

1. Simple Correlation
2. Multiple Correlation
3. Partial Correlation
4. Semi-partial Correlation

8.6.1 Simple Correlation

Let X, Y be jointly distributed as $X, Y \sim N(\mu, \Sigma)$ where

$$\mu = \begin{bmatrix} \mu_X \\ \mu_Y \end{bmatrix}$$

is the mean,

$$\Sigma = \begin{bmatrix} \sigma_{XX} & \sigma_{YY} \\ \sigma_{YX} & \sigma_{YY} \end{bmatrix}$$

is the covariance, and

$$\rho = \rho_{XY} = \rho_{YX} = \frac{\sigma_{XY}}{\sqrt{\sigma_{XX}\sigma_{YY}}}$$

is the correlation coefficient. The sample formula for the simple correlation coefficient is

$$r_{XY} = \frac{\hat{\sigma}_{XY}}{\sqrt{\hat{\sigma}_{XX}\hat{\sigma}_{YY}}} = \sqrt{\frac{\sum_i (X_i - \bar{X})(Y_i - \bar{Y})}{\sum_i (X_i - \bar{X})^2 \sum_i (Y_i - \bar{Y})^2}}$$

Example 8.15 – Simple correlation:

The simple correlation, also referred to as the "zero-order correlation," measures the linear relationship between two variables. Consider a situation in which data for two continuous variables, **X** and **Y**, are measured on each of 10 individuals in a study. The following code can be used to compute the sample correlation coefficient r_{XY} based upon the analytic formula.

```
PROC IML;
    x={3,3,4,4,5,5,6,6,7,8};
    y={3,4,4,5,5,6,6,6,7,8};
    n=NROW(x);
    r=SQRT(SSQ((y-SUM(y)/n)'*(x-SUM(x)/n))/
          ((SSQ(y-SUM(y)/n)* SSQ(x-SUM(x)/n))));
    PRINT r;
QUIT;
```

Output – Example 8.15:

```
                         r
                     0.9583828
```

8.6.2 Partial Correlation

The second type of correlation to be considered is a partial correlation. The square of the partial correlation looks at the amount of variability in a first set of variables explained by a second set of variables considering a third set of variables are held constant ("partialed" out). Consider the following diagram.

The circle designated *X1* can represent variability associated with a single variable or multiple variables. The same is true for *X2* and *X3*. The overlap in the circles represents squared correlation among the variables and/or groups of variables. The squared correlation can be described as the proportion of variability associated with each variable (or group of variables). The total area in a single circle represents all the variability associated with that variable (or group of variables). The overlap in the circles represents shared variability. For example, the area *D+E* represents the variability shared by *X1* and *X2*. The proportion,

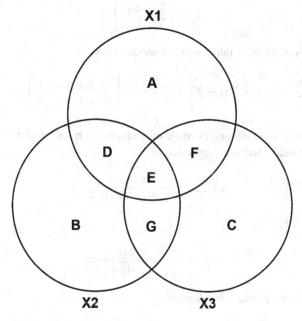

Fig. 8.2 Three-variable correlation

$$r^2_{X2X1} = \frac{D+E}{A+D+E+F}$$

represents the proportion of the total variability in $X1$ that is explained by $X2$. This proportion is the square of the simple correlation between $X1$ and $X2$.

The square of the partial correlation of $X1$ and $X2$ in $X1$ given that $X3$ is held constant is represented by the ratio

$$r^2_{12(3)} = \frac{D}{A+D}$$

By partialling out $X3$; the areas E and F are separated from $X1$ leaving only areas A and D.

Let

$$X = \begin{bmatrix} X_1 \\ X_2 \\ X_3 \\ X_4 \end{bmatrix} \sim N(\mu, \Sigma)$$

represent a column vector with four variables where

$$\Sigma = \begin{bmatrix} \Sigma_{11} & \Sigma_{12} \\ \Sigma_{21} & \Sigma_{22} \end{bmatrix},$$

such that Σ is partitioned into four submatrices. Next,

$$\Sigma_{11.2} = \Sigma_{11} - \Sigma_{12}\Sigma_{22}^{-1}\Sigma_{21} = \begin{bmatrix} \sigma_{11|(34)} & \sigma_{12|(34)} \\ \sigma_{21|(34)} & \sigma_{22|(34)} \end{bmatrix}$$

Then, the maximum likelihood estimate of the partial correlation of X_1 and X_2 given X_3 and X_4 are held fixed is computed as

$$\rho_{12|(34)} = \frac{\sigma_{12|(34)}}{\sqrt{\sigma_{11|(34)}\sigma_{22|(34)}}}$$

which is estimated by

$$r_{12|(34)} = \frac{\hat{\sigma}_{12|(34)}}{\sqrt{\hat{\sigma}_{11|(34)}\hat{\sigma}_{22|(34)}}}$$

Example 8.16 – Partial correlation:

Let

$$S = \begin{bmatrix} 285 & 240 & 205 & 180 \\ 240 & 284 & 238 & 202 \\ 205 & 238 & 281 & 234 \\ 180 & 202 & 234 & 276 \end{bmatrix}$$

be a covariance matrix estimated from a sample of data. The following code can be used to compute the partial correlation of **X1** and **X2** given **X3** and **X4** are held fixed.

```
PROC IML;
s={285 240 205 180,
   240 284 238 202,
   205 238 281 234,
   180 202 234 276};
   S11=S[1:2,1:2];
   S12=S[1:2,3:4];
   S21=S[3:4,1:2];
   S22=S[3:4,3:4];
   S11_2=S11-S12*INV(s22)*s21;
   rho=s11_2[1,2]/SQRT(s11_2[1,1]*s11_2[2,2]);
   PRINT s s11_2 rho;
QUIT;
```

Output – Example 8.16:

s				s11_2		rho
285	240	205	180	134.38158	65.934211	0.6271857
240	284	238	202	65.934211	82.241228	
205	238	281	234			
180	202	234	276			

The estimate of the partial correlation coefficient is $r_{12|(34)} = 0.627$.

8.6.3 Multiple Correlation

The multiple correlation coefficient is common in multiple regression output. One formula for the estimate of the multiple correlation coefficient is $R_{0(123)} = \sqrt{\frac{SS_{regression}}{SS_{total}}}$. It is the square root of coefficient of determination (r^2), the amount of variability in the response variable that is explained by the model. The coefficient of determination can be represented in Figure 8.2. Consider *X1* as the dependent variable and *X2* and *X3* as the independent variables, then

$$r^2 = \frac{D+E+F}{A+D+E+F}$$

is the coefficient of determination, the square of the multiple correlation.

Example 8.17 – Multiple correlation:

Using a portion of the regression analysis from Example 8.12, we can compute the maximum likelihood estimate of the multiple correlation coefficient using the following code:

```
PROC IML;
    x={71 183 11.0,69 125 9.0,71 145 11.0,69 125 8.5,
       74 180 9.0,67 135 7.0,76 185 15.0,68 120 8.5,
       66 112 7.5,66 137 9.5,64 110 8.0,66 150 8.5,
       64 100 7.5,71 196 10.5,65 120 8.5,67 125 9.0,
       66 165 9.5,67 145 8.5,72 170 11.0,70 185 8.5,
       69 190 9.0,69 186 12.0,69 145 10.0,68 150 9.5,
       66 140 8.5,67 175 10.5,68 152 9.5,67 155 8.5,
       69 150 9.5,73 175 11.0,70 170 13.0,74 192 12.0,
       71 165 10.5,61 100 7.5,68 160 10.0,72 175 12,
       62 110 6.5,64 110 7.0,68 140 10.0};
    y={136,179,188,140,164,165,178,154,186,169,179,153,
       174,186,177,179,189,196,149,181,163,130,164,192,
       170,177,179,176,180,183,164,191,172,133,165,133,
       179,183,198};
```

```
      n=NROW(x);
      x0=J(n,1);
      x=x0||x;
      b_hat=INV(x'*x)*x'*y;
      p=NCOL(x);
      resid=y - x*b_hat;
      sstot=y'*y - y'*((1/n)*J(n))*y;
      sse=y'*y - b_hat'*x'*y;
      ssr=sstot-sse;
      r2=ssr/sstot;
      mult_r = SQRT(r2);
      PRINT r2 mult_r;
QUIT;
```

Output – Example 8.17:

$$
\begin{array}{cc}
\text{r2} & \text{mult\_r} \\
0.0231063 & 0.1520075
\end{array}
$$

The maximum likelihood estimate of the multiple correlation coefficient is $R_{0(123)} = 0.1520075$.

8.6.4 Semi-partial Correlation

The squared semi-partial correlation can be explained as the additional amount of variability explained by a variable or set of variables beyond the variability that is already being explained by the remaining variables. Consider Figure 8.1. It has been explained that

$$
r_{X2X3}^2 = \frac{E+G}{C+E+F+G}
$$

represents the proportion of the total variability in *X3* that is explained by *X2*. This proportion is the square of the simple correlation between *X2* and *X3*, but can also be considered the multiple correlation where *X3* represents the dependent variable and *X2* represents the group of independent variables $\left(r_{X2X3}^2 = R_{3(2)}^2 \right)$. The proportion

$$
r_{1(3,2)}^2 = \frac{F}{C+E+F+G}
$$

represents the proportion of the total variability in *X3* that is explained by *X1* beyond that which *X2* was already explaining. The *X2* variability defined in Figure 8.1

by the letters E and G is taken out of the numerator of $r^2_{1(3,2)}$, but not out of the denominator.

The squared semi-partial correlation coefficient can be computed using the following formula:

$$r^2_{1(3,2)} = R^2_{3(12)} - R^2_{3(2)}.$$

The squared multiple correlation for a model regressing both $X1$ and $X2$ on $X3$ minus the squared multiple correlation for a model regressing only $X2$ on $X3$. The difference represents the unique amount of variability that is explained by adding the regressor(s) $X1$ to the model that consisted of regressor(s) $X2$.

Example 8.18 – Semi-partial correlation:

Consider the data from Example 8.12. The following code modifies the code from Example 8.12 to compute the semi-partial correlation (also referred to as the "part" correlation), $r_{4(0.123)}$, that partials out variables **X1** through **X3** from the independent variables (**X1** through **X4**), but not from **Y**.

```
PROC IML;
   x={71 183 11.0,69 125 9.0,71 145 11.0,69 125 8.5,
      74 180 9.0,67 135 7.0,76 185 15.0,68 120 8.5,
      66 112 7.5,66 137 9.5,64 110 8.0,66 150 8.5,
      64 100 7.5,71 196 10.5,65 120 8.5,67 125 9.0,
      66 165 9.5,67 145 8.5,72 170 11.0,70 185 8.5,
      69 190 9.0,69 186 12.0,69 145 10.0,68 150 9.5,
      66 140 8.5,67 175 10.5,68 152 9.5,67 155 8.5,
      69 150 9.5,73 175 11.0,70 170 13.0,74 192 12.0,
      71 165 10.5,61 100 7.5,68 160 10.0,72 175 12,
      62 110 6.5,64 110 7.0,68 140 10.0};
   y={136,179,188,140,164,165,178,154,186,169,179,153,
      174,186,177,179,189,196,149,181,163,130,164,192,
      170,177,179,176,180,183,164,191,172,133,165,133,
      179,183,198};
   n=NROW(x);
   x0=J(n,1);
   x=x0||x;
   START codef(x,y,n);
      b_hat=INV(x`*x)*x`*y;
      p=NCOL(x);
      sstot=y`*y - y`*((1/n)*J(n))*y;
      sse=y`*y - b_hat`*x`*y;
      ssr=sstot-sse;
      r2=ssr/sstot;
      RETURN(r2);
```

```
    FINISH;
    full=CODEF(x,y,n);
    xr=x[,1:3];
    reduced=CODEF(xr,y,n);
    diff=full-reduced;
    semipartial = SQRT(diff);
    PRINT "Semi-partial Correlation:
    unique contribution of X4",  full reduced diff semipartial;
QUIT;
```

The functional module CODEF is created to compute the coefficient of determination. The full **X** matrix is passed to CODEF and the coefficient of determination, **FULL**, is computed. Then, a reduced **X** matrix (without **X4**) is passed to CODEF and the coefficient of determination, **REDUCED**, is computed. The difference, **DIFF**, of **REDUCED** from **FULL** is computed, and the square root is taken to result in the semi-partial correlation, **SEMIPARTIAL**.

Output – Example 8.18:

```
      Semi-partial Correlation: unique contribution of X4
              full     reduced      diff semipartial
          0.0231063 0.0072309 0.0158754   0.1259975
```

As a result, the proportion of the total variability in **Y** that is uniquely described by **X4** is computed as **DIFF**=0.0158754, and the semi-partial correlation, $r^2_{4(0.123)}$, is computed as **SEMIPARTIAL**=0.1259975.

8.7 Chapter Exercises

8.1 Create a design matrix in IML using Kronecker products that represent the following fixed effects situation:

  • Two-way fixed effects (including an intercept) model with interaction.
  • The first factor, A, has 3 levels. The second factor, B, has 4 levels. This is a balanced design with 5 individuals per treatment level combination.

8.2 Create a design matrix in IML using the DESIGN function that represent the following fixed effects situation:

  • Two-way fixed means (no intercept) model with interaction.
  • The first factor, A, has 2 levels. The second factor, B, has 3 levels. This is a balanced design with 4 individuals per treatment level combination.

8.3 Create a design matrix in IML using the DESIGN function that represent the following fixed effects situation:

  • Three-way fixed effects (including an intercept) model with all interactions.

- The first factor, A, has 3 levels. The second factor, B, has 2 levels. The third level, C, has 2 levels. This is a balanced design with 3 individuals per treatment level combination.

8.4 The following data are assumed to be sample values from a bivariate normal distribution with mean μ and covariance Σ:

X	7.85	0.17	4.73	7.83	7.60	9.20	-2.55	3.30	7.29	0.74
Y	3.10	1.46	8.29	8.49	9.38	11.40	2.62	6.84	8.12	5.66

Using IML compute the LS estimates of μ and Σ.

8.5 The following data are assumed to be sample values from a bivariate normal distribution with mean μ and covariance Σ:

X	7.85	0.17	4.73	7.83	7.60	9.20	-2.55	3.30	7.29	0.74
Y	3.10	1.46	8.29	8.49	9.38	11.40	2.62	6.84	8.12	5.66

a. Compute a 95% confidence interval about μ_x.
b. Compute a 95% confidence interval about μ_y.

8.6 The following data are assumed to be sample values from a bivariate normal distribution with mean μ and covariance Σ:

X	7.85	0.17	4.73	7.83	7.60	9.20	-2.55	3.30	7.29	0.74
Y	3.10	1.46	8.29	8.49	9.38	11.40	2.62	6.84	8.12	5.66

Using IML and the method of least squares, Test the hypothesis

$$H_0:\mu_x = \mu_y \ vs. H_A:\mu_x \neq \mu_y$$

8.7 The following data are assumed to be sample values from a bivariate normal distribution with mean μ and covariance Σ:

X	7.85	0.17	4.73	7.83	7.60	9.20	-2.55	3.30	7.29	0.74
Y	3.10	1.46	8.29	8.49	9.38	11.40	2.62	6.84	8.12	5.66

Consider X and Y two treatment levels in a designed experiment with numeric responses. Using IML and the method of maximum likelihood, compute the ML estimates of μ and covariance Σ.

8.8 The following data are assumed to be sample values from a bivariate normal
distribution with mean μ and covariance Σ:

X	7.85	0.17	4.73	7.83	7.60	9.20	-2.55	3.30	7.29	0.74
Y	3.10	1.46	8.29	8.49	9.38	11.40	2.62	6.84	8.12	5.66

Fit the simple linear regression model

$$Y = \beta_0 + \beta_1 X + \varepsilon$$

using the sample data.

a. Compute the estimate of the slope and the estimate of the y-intercept.
b. Compute the residual mean squares.
c. Compute the coefficient of determination, r^2.
d. Test the hypothesis

$$H_0{:}\beta_1 = 0 \ vs. \ H_0{:}\beta_1 \neq 0$$

e. Compute a 95% confidence interval about β_1.

8.9 The following code generates data from a multivariate normal distribution with
mean

$$\mu = \begin{bmatrix} 3 \\ 5 \\ 7 \end{bmatrix}$$

and covariance

$$\Sigma = \begin{bmatrix} 5 & 2 & 4 \\ 2 & 8 & 2 \\ 4 & 2 & 5 \end{bmatrix}.$$

```
PROC IML;
    mu={3,5,7};
    sigma={5 2 4,
           2 8 2,
           4 2 5};
    n=100;
    CALL VNORMAL(y,mu,sigma,n);
QUIT;
```

a. Compute the simple correlation coefficients, r_{12}, r_{13}, and r_{23}.
b. Compute the multiple correlation, $R_{1(23)}$, $R_{2(23)}$, and $R_{3(12)}$.
c. Compute the partial correlations, $r_{12(3)}$, $r_{13(2)}$, and $r_{23(1)}$.
d. Compute the semi-partial (part) correlations, $r_{1(3.2)}$, $r_{1(2.3)}$, $r_{2(1.3)}$, $r_{2(3.1)}$, $r_{3(1.2)}$, and $r_{3(2.1)}$.
e. Identify any answers to to any of the values computed for Parts a through d that yield equivalent numeric values. Explain why those values are the same.

Chapter 9
Linear Mixed Models

9.1 Mixed Models Theory

In the early 1950s, C.R. Henderson developed mixed model estimation, something
he began in the 1940s with his Ph.D. thesis. He wanted to analyze data for a linear
model with fixed environmental and random genetic factors in the breeding of swine
(Van Vleck, 1998). This mixed model would allow for both fixed and random effects
to be included in the same model. The mixed model allows for the incorporation of
information about the correlation or covariance structure of data that is essentially
disregarded if the assumption of independence is employed. There are many applications for the mixed model and mixed model analyses are now very common in
applied research.

The mixed model is defined by Wolfinger, et. al. (1991) with the following form:

$$Y = X\beta + Zu + e \tag{9.1}$$

where y represents the observed responses, X is the design matrix for the model's
fixed effects, β is the vector of fixed effects, Z is the design matrix for the model's
random effects, u is the vector of random effects, and e is the vector of unknown
random errors. It is assumed that both u and e are normally distributed with

$$E\begin{bmatrix} u \\ e \end{bmatrix} = \begin{bmatrix} 0 \\ 0 \end{bmatrix}$$

and

$$Var\begin{bmatrix} U \\ e \end{bmatrix} = \begin{bmatrix} G & 0 \\ 0 & R \end{bmatrix}$$

where G represents the variance of u and R represents the variance of e. Note that
with the non-diagonal elements of the above variance matrix equaling zero, the
implicit assumption is that u and e are independent. It is also assumed that both
G and R are nonsingular. The variance of Y can then be computed as

J.J. Perrett, *A SAS/IML Companion for Linear Models*, Statistics and Computing,
DOI 10.1007/978-1-4419-5557-9_9, © Springer Science+Business Media, LLC 2010

$$V = Var(Y) = ZGZ' + R. \tag{9.2}$$

where Z represents the design matrix for the random effects, much as X represents a design matrix for the fixed effects. If u represents a vector of zeros and R has the form $\sigma^2 I$ then the mixed model reduces to the standard linear model.

There is not a least squares solution to the mixed model equations as with the standard linear model, due to the number and complexity of unknown parameters. Instead, maximum likelihood estimation is used to estimate the unknown parameters, conditional upon the random effects having a specific design structure. Because there are many possible design structures associated with a given set of data, one approach to analyzing mixed model data is to determine an appropriate covariance structure that approximates the process that resulted in the obtained data, and then to estimate the set of unknown parameters and predict random variables, conditional upon the chosen covariance structure, that maximize the likelihood associated with the mixed model and the collected data. One can choose multiple candidate covariance structures and compare them based on fit statistics. Once a covariance structure is ultimately chosen, hypothesis tests for fixed effects can be performed based on the model incorporating the chosen covariance structure.

The SAS System includes a PROC step, PROC MIXED, that analyzes data that can be modeled with both fixed and random effects. This section will demonstrate mixed model analysis using SAS/IML. The results will be compared to the results of the mixed model analysis using PROC MIXED, which will be considered the standard.

If the matrices G, R, and consequently V are known matrices of constants, then the following mixed model equations (Henderson 1984)

$$\begin{bmatrix} X'R^{-1}X & X'R^{-1}Z \\ Z'R^{-1}X & Z'^{R^{-1}}Z + G^{-1} \end{bmatrix} \begin{bmatrix} \hat{\beta} \\ \hat{u} \end{bmatrix} = \begin{bmatrix} X'R^{-1}Y \\ Z'R^{-1}Y \end{bmatrix}$$

can be solved for the estimates of both B and U as

$$\hat{\beta} = (X'V^{-1}X)^{-1}X'V^{-1}Y$$

$$\hat{u} = GZ'V^{-1}(Y - X\hat{\beta})$$

respectively. However, in practice, V, R, and G will be replaced by estimates $\hat{V}$, $\hat{R}$, and $\hat{G}$. These estimates can then be used for testing hypotheses and constructing confidence intervals using estimable linear combinations of $\hat{\beta}$ and $\hat{u}$.

9.2 Mixed Model Analysis

The analysis of a mixed model is a process containing several steps. Part of that process involves finding maximum likelihood estimates. In the case of mixed models, restricted maximum likelihood (REML) estimation may be used. Maximum likelihood estimation for obtaining estimates of covariance parameters leads to underestimation of the variance on average, thought that bias decreases as the sample size increases. With REML, likelihood estimation is performed on the data after it has been transformed. The transformation allows for estimation of the covariance parameters, profiling out the fixed-factor parameters. The resulting REML estimates of the variance are unbiased.

Consider a situation with three factors: A is a fixed factor, B is a random factor, and the AB interaction is also a random factor. Factor A might represent the specifically chosen treatments being applied to patients in a multi-site clinical trial and B might represent the randomly selected sites. The following matrices will be defined or computed as part of the mixed models analysis process:

- X: The design matrix for the fixed effects. Depending on the number of fixed effects, X will be a $n \times (p+1)$ matrix of rank $= p$, where n represents the number of observed responses and p is the total number of factor levels and/or continuous variables of the fixed effects. The addition of 1 in the $n \times (p+1)$ is to account for a column of ones in the far left column of the matrix X to represent the intercept or mean of Y. For example, with A fixed, B random, and AB random, only the factor A is designed in the X matrix. In this case, if A has a factor levels, X will have n rows and $a + 1$ columns. For an alternate example, if there had been an additional fixed factor C with c levels and a fixed AC interaction with $a \times c$ levels then $p = a + c + a \times c$, to account for the total number of levels of the fixed effects. Or, if C were a continuous variable then C counts as one variable, and AC has a $\times$ 1 levels, $p = a + 1 + a \times 1 = 2a + 1$.
- M: The projection matrix—the projection onto x-space. $M = I_n - X(X'X)^{-1}X'$
- $\hat{C}$: An estimate of the MP-inverse of the coefficient matrix in the mixed model equations.

$$\hat{C} = \begin{bmatrix} X'\hat{R}^{-1}X & X'\hat{R}^{-1}Z \\ Z'\hat{R}^{-1}X & Z'\hat{R}^{-1}Z + \hat{G}^{-1} \end{bmatrix}^{-}$$

- Z_1: A matrix consisting of the columns of the matrix Z that apply to the random Factor B.
- Z_2: A matrix consisting of the columns of the matrix Z that apply to the random Factor AB interaction.
- Z: The design matrix for the random effects. Depending on the number of random effects, Z will be a $n \times q$ matrix of rank q. It can be computed as the horizontal concatenation of the matrices Z_1 and Z_2. $Z = Z_1||Z_2$
- $\hat{G}_1 = \hat{\sigma}_B^2 I_b$, where $\hat{\sigma}_B^2$ represents the estimate of the variance component associated with Factor B.
- $\hat{G}_2 = \hat{\sigma}_{AB}^2 I_{ab}$, where $\hat{\sigma}_{AB}^2$ represents the estimate of the variance component associated with Factor AB interaction.
- $\hat{G}$: The estimated variance of $\hat{u}$. $\hat{G}$ is a block-diagonal matrix of the blocks $\hat{G}_1$ and $\hat{G}_2$.
- $\hat{R}$: The estimated variance of e. $\hat{R} = \tilde{\sigma}_s^2 I_n$, where $\tilde{\sigma}_s^2$ represents the estimate of the variance component associated with the random error.

- $\hat{V}$: The covariance matrix. The matrix $\hat{V}$ is computed as $\hat{V} = Z\hat{G}Z' + \hat{R}$ but, can also be computed in terms of the individual components of $\hat{G}$ as $\hat{V} = Z_1\hat{G}_1Z'_1 + Z_2\hat{G}_2Z'_2 + \hat{R}$.

The process of obtaining estimates for the mixed model are outlined as follows: First, a Newton-Raphson Ridge-stabilized optimization routine is used to obtain values for the covariance parameters. The optimization is actually found by minimizing -2 multiplied by the log of the REML equation:

$$-2l_R(\theta;X) = log\,|V(\theta)| + log\,|X^*[V(\theta)]^{-1}X^*| + r'[V(\theta)]^{-1}r + (n-a)\,log\,(2\pi)$$

where θ refers to the covariance parameters made up of all unknowns in the G and R matrices (Littell et. al., 2006); X^* represents the nonsingular version of X, without the initial column of ones; and r is defined as

$$r = Y - X^*(X^*[V(\theta)]^{-1}X^*)^{-1}X^{*'}[V(\theta)]^{-1}Y.$$

Taking the log of the REML equation and multiplying by -2 reduces the order of the equation requiring less computer-intensive computation for its solution. It also requires that the optimization routine seek to *minimize* the equation rather than maximize it. Once the optimization routine has found the value of θ, say $\hat{\theta}_R$, that minimizes -2 multiplied by the log of the REML equation, that value, $\hat{\theta}_R$, is then used to compute $\hat{R}$ and $\hat{G}$. The following values are then computed:

$$\hat{V} = Z\hat{G}Z' + \hat{R}$$

which is then used to compute

$$\hat{\beta} = (X'\hat{V}^{-1}X)^{-1}X'\hat{V}^{-1}Y \tag{9.3}$$

$$\hat{u} = \hat{G}Z'\hat{V}^{-1}(Y - X\hat{\beta}) \tag{9.4}$$

and

$$\hat{C} = \begin{bmatrix} X'\hat{R}^{-1}X & X'\hat{R}^{-1}Z \\ Z'\hat{R}^{-1}X & Z'\hat{R}^{-1}Z + \hat{G}^{-1} \end{bmatrix}^{-} \tag{9.5}$$

A more efficient way to compute the estimates $\hat{\beta}$ and $\hat{u}$ is by using the Sweep method (Explained with more details in Section 10.2). Using the Sweep method, first construct

$$\hat{M} = \begin{bmatrix} X'\hat{R}^{-1}X & X'\hat{R}^{-1}Z & X'\hat{R}^{-1}Y \\ Z'\hat{R}^{-1}X & Z'\hat{R}^{-1}Z + \hat{G}^{-1} & Z'\hat{R}^{-1}Y \\ Y'\hat{R}^{-1}X & Y'\hat{R}^{-1}Z & Y'\hat{R}^{-1}Y \end{bmatrix}. \tag{9.6}$$

Then,

$$
SWEEP(\hat{M}, \hat{C}^-) = \begin{bmatrix} \hat{C} & \begin{bmatrix} \hat{\beta} \\ \hat{u} \end{bmatrix} \\ -[\hat{\beta}\ \hat{u}] & (Y - X\hat{\beta})'\hat{V}^{-1}(Y - X\hat{\beta}) \end{bmatrix} \tag{9.7}
$$

yielding the estimates $\hat{\beta}$ and $\hat{u}$ without having to invert the matrix V. Had the matrices G and R been known, the estimates for β and u would have been best linear unbiased estimator (BLUE) and best linear unbiased predictor (BLUP), respectively. However, because estimates of the matrices G and R were used to compute $\hat{\beta}$ and $\hat{u}$, they are considered estimated (or empirical) best linear unbiased estimator (EBLUE) and estimated (or empirical) best linear unbiased predictor (EBLUP), respectively.

Linear functions of the estimates can be used for constructing hypothesis tests and confidence intervals. Estimable linear combinations of the form

$$
L \begin{bmatrix} \beta \\ u \end{bmatrix}
$$

have estimates of the form

$$
L \begin{bmatrix} \hat{\beta} \\ \hat{u} \end{bmatrix}.
$$

When L contains one row, the hypothesis

$$
H_0 : L \begin{bmatrix} \beta \\ u \end{bmatrix} = 0 \tag{9.8}
$$

is tested using the test statistic

$$
t = \frac{L \begin{bmatrix} \hat{\beta} \\ \hat{u} \end{bmatrix}}{\sqrt{L\hat{C}L'}} \tag{9.9}
$$

compared to a t value with υ degrees of freedom. The degrees of freedom are approximated using a Satterthwaite, Kenward-Rodgers, or other method.
A $(1 - \alpha) \times 100\%$ confidence interval about

$$
L \begin{bmatrix} \beta \\ u \end{bmatrix}
$$

is computed using

$$
L \begin{bmatrix} \hat{\beta} \\ \hat{u} \end{bmatrix} \pm t_{\alpha/2,v} \sqrt{L\hat{C}L'} \tag{9.10}
$$

where $t_{\alpha/2,v}$ is a value from the t-distribution with tail probability $\alpha/2$ and approximated degrees of freedom v.

When L contains more than one row, relabeled as H, the hypothesis

$$H_0 : H \begin{bmatrix} \beta \\ u \end{bmatrix} = 0 \qquad (9.12)$$

is tested using the test statistic

$$F = \frac{\left(H \begin{bmatrix} \hat{\beta} \\ \hat{u} \end{bmatrix} \right)' \left[H\hat{C}H' \right]^{-1} \left(H \begin{bmatrix} \hat{\beta} \\ \hat{u} \end{bmatrix} \right)}{rank(H)} \qquad (9.13)$$

compared to an F value with numerator degrees of freedom equal to *rank* (*H*) and denominator (approximated) degrees of freedom v. Simultaneous confidence intervals can be computed using the formula

$$H \begin{bmatrix} \hat{\beta} \\ \hat{u} \end{bmatrix} \pm t_{\alpha/2:nh,v} \left[H\hat{C}H \right]^{1/2} \qquad (9.14)$$

where $t_{\alpha/2:nh,v}$ represents a quantile from the multivariate-t distribution with probability $\alpha/2$, *nh* equal to the number of linearly independent estimable functions in *H*, and degrees of freedom v. This value can be computed using the SAS function PROBMC('MAXMOD',.,prob,df,nh).

Example 9.1 – Mixed model analysis using IML:

The following SAS code generates a set of mixed model data for analysis using PROC IML. The mixed model for this example can be defined as follows:

$$y_{ijk} = \mu + \alpha_i + b_j + \alpha b_{ij} + e_{ijk}, i = 1\ldots4, j = 1\ldots5, k = 1\ldots3 \qquad (9.15)$$

where

$$b_j \sim Normal(0, \sigma_b^2), \quad \sigma_b^2 = 25$$

$$\alpha b_{ij} \sim Normal(0, \sigma_{ab}^2), \quad \sigma_{ab}^2 = 4$$

$$eb_{ijk} \sim Normal(0, \sigma_e^2), \quad \sigma_e^2 = 1$$

Because of the large amount of code for this example, the code, and notably the PROC IML step, will be divided into smaller sections, as will the output. Each section will be preceded by an explanation.

The first DATA step generates data for a two factor and interaction mixed model with one fixed factor, A, and random factors B and AB. The data are generated to help illustrate the mixed model analysis. The variables **A**, **B**, and **C** indicate the number of levels of Factors A, B, and the number of reps, respectively. The variables **EB** and **EAB** represent the variance of Factors B and the AB interaction, respectively. The variable **EB** is formulated such that the underlying distribution is Normal with mean $= 0$ and standard deviation $= 5$ (i.e.- variance $= 25$). The variable **EAB** is formulated such that the underlying distribution is Normal with mean $= 0$ and standard deviation $= 2$ (i.e.- variance $= 4$). The random error (sampling error) is generated such that the underlying distribution is Normal with mean $= 0$ and standard deviation $=$ variance $= 1$. The DO loops are used to generate each data value, y_{ijk}, based on the underlying mixed model as defined in Equation 9.15. The expected value of Y is defined as **Y**=10 for Factor A level 1, **Y**=20 for level 2, **Y**=30 for level 3, and **Y**=40 for level 4. The expected value of Y is coded using 10***I** in the assignment statement for **Y** in the DATA step. The variable **I** represents the level of Factor A, and multiplying by 10 produces the expected value of **Y**=10, 20, 30, or 40. Giving each level of the fixed factor, A, a different expectation leads to significant results in the hypothesis test of the effect of Factor A in the analysis. The data set DSET1 is generated as a result of the DATA step, containing the factor levels and the responses.

The second DATA step was added to make the data in DSET1 unbalanced. This was done for illustrative purposes to add complexity to the analysis.

```
DATA dset1;
   a = 4;
   b = 5;
   c = 3;
   DO j = 1 TO b;
      eb = RANNOR (0)*5; /** Variance of random effect
           of factor B **/
      DO i = 1 TO a;
         eab = RANNOR (0)*2; /** Variance of random effect of
              A*B interaction **/
         DO k = 1 TO c;
            y = 10*i + eb + eab + RANNOR (0); /** Generation
                 of response vector **/
            OUTPUT;
         END;
      END;
   END;
RUN;

DATA dset1;
   SET dset1;
   IF i=1 AND j=1 THEN DELETE;
RUN;
```

The PROC SORT step orders the data by increasing factor. This aids in the creation of the design matrices in the analysis.

```
PROC SORT DATA=dset1;
   BY i j k;
RUN;
```

The PROC IML step is used to analyze the mixed model data in the SAS data set DSET1. The response data, variable **Y**, are read into the 60×1 column vector **Y**. The factor A level data are read into the 60×1 column vector **X1**. The Factor B level data are read into the 60×1 column vector **Z**. The variable **B** (which has a value of 5 for all 60 rows) data are read into the 60×1 column vector **B1**, and is then assigned to the scalar variable, **B**, with the one value, 5, as opposed to 60 rows with the same value.

The RESET FUZZ; statement formats the output such that numeric results with numbers smaller than 1E-12 (due to finite precision arithmetic) are set to zero, rather than being displayed in scientific notation.

The matrix **XNS** is the full column-rank (nonsingular) design matrix for the fixed effects. The matrix **X** is created by appending a column of ones to left of the matrix **X1**. The rank of the design matrix, **X**, is assigned to the variable **A**. The design matrix for the random effect of Factor B is created using the DESIGN function on the column vector **Z** and assigned to the matrix **Z1**. The design matrix for the random effect of the AB interaction is created using the DESIGN function on the mathematical function

$$b(x_1 - 1) + z,$$

which generates a unique value for each combination of the levels of Factor A and the levels of Factor B. This function produces the necessary setup of ones and zeros for creating the appropriate design matrix. The modular function MATRANK is created to compute matrix ranks throughout the program. The number of rows in the design matrix, **X**, is assigned to the variable **N**.

```
PROC IML;
   USE dset1;
   READ ALL VAR {y} INTO y;
   READ ALL VAR {i} INTO x1;
   READ ALL VAR {j} INTO z;
   READ ALL VAR {b} INTO b1;
   RESET FUZZ;
   xns=DESIGN (x1); /** non-singular (full column rank)
   design matrix for fixed effects **/
   x=J (NROW (xns),1,1)||xns; /** singular
   (over parameterized) design matrix for fixed effects **/
   START matrank (x); /** matrank computes matrix rank **/
      matrank=ROUND (TRACE (x*GINV (x)));
      RETURN (matrank);
```

```
FINISH;
a=matrank (x);
b=b1 [1,1];
z1=DESIGN (z);
z2=DESIGN (b*(x1-1)+z); /** z1 and z2 make up design
    matrix for random effects **/
n = NROW (x);
```

The identity matrices; **I11**, **I22**, and **I33**; are set up to accommodate portions of the covariance structure of the model. The variable **PI** is assigned the value of *PI* (i.e.- 3.1415...) using the CONSTANT function.

```
i11 = I (NCOL (Z1)); /** dimension=b when balanced **/
i22 =I (NCOL (Z2)); /** dimension=a*b when balanced **/
i33 =I (n);
pi = CONSTANT (pi = CONSTANT ("PI");
```

The REML function is created as a module. It's purpose is to define the function of the log of the REML equation multiplied by -2 (written as *-2logREML*). The function requires the value of the 1×3 row vector theta to be passed and retains as global values, **Z1**, **Z2**, **Y**, **XNS**, **I11**, **I22**, **I33**, **PI**, **N**, and **A**. These values are defined prior to the start of the REML module so that when the module is called multiple times the global values need not be computed multiple times. Defining the global values within the module would then cause them to be defined each time the module is called. The 1×3 row vector **THETA** is the vector of variance components pertaining to the mixed model. The three elements represent the input values of $\sigma_b^2, \sigma_{ab}^2$, *and* σ_e^2, respectively. The matrices **SIGMA_B**, **SIGMA_AB**, and **SIGMA_E** are assigned the input variance component values multiplied by the appropriate-size identity matrix. The covariance matrix, **V**, is then defined using the variance components and their associated design matrices. The variable, **PROD**, is the result of the function *-2logREML*. The

```
RETURN (prod);
```

statement indicates that the module is acting as a function and will return the value of prod as the function result.

```
START reml (theta) GLOBAL (z1,z2,y,xns,i11,i22,i33,pi,n,a);
/** Defining the function -2 (Log Likelihood) **/
    sigma_b = theta[1, 1]*i11;
    sigma_ab = theta[1, 2]*i22;
    sigma_e = theta[1, 3]*i33;
    v = z1*sigma_b*z1' +z2*sigma_ab*z2' +sigma_e;
    /** Covariance matrix **/
    vinv=INV (v);
    r=y-xns*GINV(xns'*vinv*xns)*xns'*vinv*y;
    prod=LOG (DET(v))+LOG (DET (xns'*vinv*xns))+
    r'*vinv*r+(n-a)*LOG(2*pi);
```

```
    RETURN (prod);
 FINISH reml;
```

Next, the minimum variance quadratic unbiased estimates, MIVQUE(0), of the variance components are computed. When the data are balanced and the covariance structure is basic, the MIVQUE(0) estimates are the REML estimates. Consequently, PROC MIXED first computes the MIVQUE(0) estimates. If further estimation is required due to a lack of balance in the data or due to a more complex covariance structure like a spatial structure, the MIVQUE(0) estimates are used as the starting values for the minimization routine. For this example, the data were intentionally designed as unbalanced so that the minimization routine would be required.

```
/** MIVQUE (0) **/
   p=i33-x*GINV(x); /** projection onto x **/;
   pz1=p*z1*z1';
   pz2=p*z2*z2';
   pz3=p*i33;
   mat_a=(TRACE (pz1*pz1)||trace (pz1*pz2)||TRACE (pz1*pz3))//
         (TRACE (pz2*pz1)||trace (pz2*pz2)||TRACE (pz2*pz3))//
         (TRACE (pz3*pz1)||trace (pz3*pz2)||TRACE (pz3*pz3));
   mat_b=(y'*pz1*p'*y)//(y'*pz2*p*y)//(y'*pz3*p*y);
   mivque0=(GINV(mat_a)*mat_b)';
   print mivque0;
```

The next part of the program is the minimization of *-2logREML*. First, the inputs for the minimization routine are defined. The 1×3 row vector **MIVQUE0** contains the initial estimates of the variance components. These initial estimates need to be close to the population values in order for the optimization routine to converge at reasonable estimates.

The 1×2 row vector **OPTN** specifies some of the options for the optimization routine. The first option indicates whether maximization or minimization will be utilized in the routine. The value 0 is the default and indicates maximization. This example requires minimization and so the value 1 is given. The second option indicates the amount of output that will be displayed as a result of the optimization routine. Smaller numbers provide less output. There are other options for the optimization routine that could have been specified. When not specified, options resort to their default values.

The 1×3 row vector **CON** specifies the constraints on the estimates that result from the optimization routine. The three estimates produced for this example are all variance components. Therefore, it makes sense to constrain these values to be positive. The value zero as a constraint, constrains the estimates to be greater than zero.

The NLPNRR call is the call statement for the Ridge-stabilized Newton-Raphson Nonlinear Optimization routine. This routine was chosen for this example because it is the routine used by PROC MIXED for REML estimation. The NLPNRR call is passed the following values: **RC**, **COVS**, REML, **MIVQUE0**, **OPTN**, and **CON**. The value for **RC**, the return code, is one of the results of the routine. The return code is a number indicating why the routine stopped. Positive return codes (1 through 10) indicate positive outcomes: absolute or relative convergence. Negative return codes (-1 through -10) indicate negative outcomes: objective function or derivative cannot be evaluated, problems with constraints, etc. The list of return codes can be found by searching the text "definition of return codes" in the online help search engine. The value **COVS** contains the 1×3 row vector of covariance parameter estimates resulting from the optimization routine. The input REML is the function to be minimized by the optimization routine. The other three input values **MIVQUE0**, **OPTN**, and **CON** are the initial values, routine options, and constraints, respectively described previously.

```
optn = {0 1}; /** First option indicates minimization,
                second option indicates amount of output **/
con={ 0 0 0 }; /** Constrains parms >= 0 **/
CALL NLPNRR (rc,covs, "reml", mivque0,optn,con); /** Ridge-
stabilized Newton-Raphson optimization routine **/
PRINT covs; /** Print the covariance parameter estimates. **/
```

The matrix **R** is assigned the error variance matrix. The matrices **G1** and **G2** are assigned the variance matrices for the random effects of Factor B and the Factor AB interaction, respectively. The matrix **G** is assigned the covariance matrix for the random effects of Factor B and the Factor AB interaction. The matrix **G** is a block-diagonal matrix created using the BLOCK function on the matrices **G1** and **G2**. In general, $G = var(u)$. The matrix **Z** is the design matrix for the random effects of Factor B and the Factor AB interaction. The matrix **V**, the variance of the mixed model, and is computed according to Equation 9.2.

```
g1=covs [1,1]*i11;
g2=covs [1,2]*i22;
g=BLOCK (g1,g2);
z=z1||z2;
r=covs [1,3]*i33;
v=z*g*z'+r;
```

The column vectors **BETA_HAT** and **U_HAT** contain the estimates of the fixed effects and the random effects using Equations 9.3 and 9.4, respectively. The matrix **C** is an estimate of the MP-inverse of the coefficient matrix in the mixed model equations and is computed according to Equation 9.5. The matrix **C** is used to determine the standard error of the estimate in hypothesis tests and confidence intervals involving linear combinations of the fixed effects parameters. The MP-inverse of **C**, **CINV**, is used in conjunction with the SWEEP function (see Equation 9.7) to compute more accurate estimates of the fixed effects and predictors of the random effects. The matrix **M** is an augmented matrix that includes the **CINV** matrix. The column vector **EST** contains the estimates of the fixed effects and predictors of the random effects and is computed more efficiently than by using Equations 9.3 and 9.4 because it does not require inverting the **V** matrix, which can lead to finite precision arithmetic errors.

```
beta_hat=GINV(x'*INV (v)*x)*x'*INV (v)*y;
u_hat=g*z'*INV (v)*(y-x*beta_hat);
PRINT beta_hat u_hat;
rinv=INV (r);
c=GINV(((x'*rinv*x)||(x'*rinv*z))//
      ((z'*rinv*x)||(z'*rinv*z+INV(g))));
cinv=GINV (c);
M=(cinv||((x'*rinv*y)//(z'*rinv*y)))//
   ((y'*rinv*x)||(y'*rinv*z)||(y'*rinv*y));
sweepmc=SWEEP (m,1:NROW(cinv));
est=sweepmc [1:NROW(cinv),NCOL(cinv)+1]; ** Estimates of
fixed effects and predictors of random effects;
PRINT est;
```

The 1×2 row vector **L** $(L = [0\ 1\ -1])$ contains contrast coefficients used both to estimate Factor level A_1 minus Factor level A_2 for construction of a confidence interval about that difference, but also to test the hypothesis that mean for Factor level A_1 is different than the mean for Factor level A_2. The second assignment of **L**,

```
l=l||J (1,NROW (est)-NCOL (l),0);
```

is used to expand the 1×2 row vector **L** $(L = [0\ 1\ -1])$ to a $1 \times$ (# *of estimates and predictors*) row vector, appending zeros as coefficients for all other estimates and predictors in the **EST** row vector. The module, MIXED_EST, is used to check **L** for estimability. The scalars **ESTIMATE** and **STDERR** contain the estimate of the linear function tested in Equation 9.8 along with its standard error. The test statistic, **T**, represents Equation 9.9 and tests the hypothesis represented in Equation 9.8. Degrees of freedom, **DF**, are computed using a formula that accounts for the unbalance in the data similar to the *containment method* used as a default in MIXED. The P value, **P_VALUE**, for the test is constructed using the PROBT function and is then formatted (**PVAL**) to have an appearance similar to that of P values found in the output of PROC MIXED. The test statistic, degrees of freedom, and P value are then printed to the output.

Using Equation 9.10 a confidence interval about the linear contrast tested in Equation 9.8 can be constructed using methods similar to those used to construct the test statistic, **T**. This exercise will be left to the reader.

```
l={0 1 -1};
l=l||J (1,NROW (est)-NCOL (l),0);
/** Create Module: mixed_est to check for estimability **/
START mixed_est(x,z,l);
   l1=l[1:NCOL(x),];
   l2=l[NCOL(x)+1:NROW(l),];
```

```
xl=x'||11;
zl=z'||12;
IF matrank(xl)^= matrank(x) | matrank(11)^= NCOL(11) |
    matrank(zl)^= matrank(z) | (matrank(12)^= NCOL(12) &
    matrank(12)^= 0)
THEN PRINT "Warning: function" 1 "is not estimable!";
FINISH;

/** check for estimability of 1 **/
CALL mixed_est(x,z,1');

estimate=1*est;
stderr=sqrt(1*c*1');
t=estimate/stderr;
rankz2=matrank(z2);
df=(a-1)*(b-1)-(a*b-matrank(Z2));
p_value=2*(1-PROBT (ABS(t),df));
IF p_value<.0001 THEN pval="<.0001";
ELSE pval=p_value;
PRINT estimate stderr t df pval;
```

The 3×5 matrix **H** contains contrast coefficients used to test for Factor A significance, or the difference in the mean response for the different levels of Factor A. It simultaneously tests the contrasts represented by the contrast coefficients in each of the rows of the matrix **H**. The second assignment of **H**,

```
h=h||J (NROW (h),NROW (est)-NCOL (h),0);
```

is used to expand the 3×5 matrix **H** to a $3 \times$ (# *of estimates and predictors* + 1) matrix, appending zeros as coefficients for all other estimates and predictors in the **EST** row vector (as was done previously with the row vector **L**). The module, MIXED_EST, is used to check **H** for estimability/testability. The test statistic, **F**, represents Equation 9.13 and tests the hypothesis represented in Equation 9.12. The numerator degrees of freedom are computed as the rank of the **H** matrix and the denominator degrees of freedom are the same as they were for the t-test described previously. The *P* value is based on the F distribution, computed using the function PROBT and then formatted as with the *P* value associated with the t-test described previously. A label scalar called **EFFECT** is created to label the output and is given the value "Factor A". The **EFFECT** label, test statistic, degrees of freedom, and *P* value are then printed to the output in a form similar to that produced by PROC MIXED.

Simultaneous confidence intervals about the linear contrasts tested in Equation 9.12 can be constructed using Equation 9.15. This exercise is left to the reader.

```
h={0 1 -1 0 0,0 1 0 -1 0,0 1 0 0 -1};
h=h||J (NROW (h),NROW (est)-NCOL (h),0);
/** check for estimability of h **/
```

```
CALL mixed_est(x,z,h');

rankh=matrank (h);
f=((est'*h')*INV (h*c*h')*(h*est))/rankh;
df1=rankh; /** compute degrees of freedom **/
df2=df;
p_value=1-PROBF (f,df1,df2);
IF p_value<.0001 THEN pval="<.0001";
ELSE pval=p_value;
effect={"Factor A"};
PRINT "Type 3 Tests of Fixed Effects",
      effect f df1 df2 pval;
QUIT;
```

The following is only selected output, key to explanation of the procedure.

Output – Example 9.1:

	mivque0	
19.847161	2.284075	1.3087335

functn
220.9143

The first section of output includes the MIVQUE(0) estimates for the variance components. These values are used as initial values for the optimization routine. If the data were balanced in this example, MIXED would skip the optimization routine and would use the MIVQUE(0) estimates for the analysis.

Output – Example 9.1: (continued):

Newton-Raphson Ridge Optimization
Without Parameter Scaling
Gradient Computed by Finite Differences
CRP Jacobian Computed by Finite Differences

Parameter Estimates	3
Lower Bounds	3
Upper Bounds	0

			Optimization Start				
Active Constraints			0		Objective Function		220.91430259
Max Abs Gradient Element			1.9121517514				

Iteration	Restarts	Function Calls	Active Constraints	Objective Function	Objective Function Change	Max Abs Gradient Element	Ridge	Ratio Between Actual and Predicted Change
1	0	2	0	220.11948	0.7948	0.6041	0	1.211
2	0	3	0	219.96055	0.1589	0.1263	0	1.145
3	0	4	0	219.95027	0.0103	0.00939	0	1.050
4	0	5	0	219.95021	0.000064	0.000063	0	1.004
5	0	6	0	219.95021	2.877E-9	1.652E-6	0	0.982

	Optimization Results		
Iterations	5	Function Calls	7
Hessian Calls	6	Active Constraints	0
Objective Function	219.95020819	Max Abs Gradient Element	1.65229E-6
Ridge	0	Actual Over Pred Change	0.9821761324

GCONV convergence criterion satisfied.

The above section of output contains the iteration history for the optimization routine. The objective function listed as the "Optimization Start" has a value of 220.91430259 and is the value of the $-2logREML$ equation with the MIVQUE(0) estimates. At each step of the iteration, the value of the objective function matches that of MIXED. Though both MIXED and this IML example use a ridge-modified Newton-Raphson optimization routine, this IML method uses finite differencing to obtain the gradient and the Hessian matrix, where MIXED computes the analytic first and second derivatives.

Output – Example 9.1: (continued):

	covs	
17.575961	3.5702504	1.3087336

beta_hat	u_hat
20.46888	-1.116721
-10.15075	-6.447087
1.0256675	1.6783259
10.241417	4.2216948
19.352545	1.6637872
	0.8955228
	-0.284216
	-1.523283
	0.9119755
	-0.433207
	-1.842942
	0.6446069
	3.4604485
	-1.828905
	-1.431382
	1.4651727
	-0.521724
	-0.173926
	0.6618591
	1.6377475
	-1.827367
	0.502255
	-0.905676
	0.5930403

est
39.821425
-29.50329
-18.32688
-9.111128
0
-1.116721
-6.447087
1.6783259
4.2216948
1.6637872
0.8955228
-0.284216
-1.523283
0.9119755
-0.433207
-1.842942
0.6446069
3.4604485
-1.828905
-1.431382
1.4651727

est
-0.521724
-0.173926
0.6618591
1.6377475
-1.827367
0.502255
-0.905676
0.5930403

In the above section of output, the vector **COVS** contains the final estimates of the variance components determined by the optimization routine. The **BETA_HAT** and **U_HAT** vectors contain the estimates of the fixed effects and predictors of the random effects based on Equations 9.3 and 9.4. The **EST** vector contains the estimates of the fixed effects and predictors of the random effects using the SWEEP method (Equation 9.7)—as is used in MIXED. The estimates in **BETA_HAT** differ from those in **EST** and MIXED due to estimation restrictions. Adding the estimate of the intercept to the other estimates in BETA_HAT yields the same results as adding the estimate of the intercept to the other estimates in EST. For example, adding 20.46888 and -10.15075 yields 10.31813 as does adding 39.821425 and -29.50329 (with rounding error).

Output – Example 9.1: (continued):

estimate	stderr	t	df	pval
-11.17642	1.3656435	-8.183993	11	<.0001

Type 3 Tests of Fixed Effects				
effect	f	df1	df2	Pval
Factor A	173.66457	3	11	<.0001

The above section of output contains output for the estimate of the difference of Factor level A_1 minus Factor level A_2, the standard error of that difference, and the test of that difference being equal to zero (Equation 9.8). That output is comparable to the output created testing the same contrast using the ESTIMATE statement in MIXED. Below that is a table of the output for the test of the effect of Factor A main effect (Equation 9.12) and is formatted to look similar to MIXED output.

Example 9.2 – Mixed model analysis using MIXED:

The following code conducts a similar analysis to that performed by the prior PROC IML step in the previous example. The CLASS statement identifies the class (or categorical) variables. The NOPROFILE option requests that the residual error variance component not be profiled out of the *-2logREML* during the optimization routine as is the default. The MODEL statement identifies the response variable and the variable representing the fixed effect, and defines the relationship between the two. The MIXED procedure uses the MODEL statement to inform the creation of the design matrix, $\mathbf{X}$ (from Equation 9.1), for the fixed effects. The SOLUTION option requests that the estimates of the factor effects be printed to the output. The random statement lists the random factor effects for the model. The MIXED procedure uses the RANDOM statement to inform the creation of the design matrix, $\mathbf{Z}$ (from Equation 9.1), for the random effects. The CONTRAST statement informs the creation of the contrast coefficient vector $\mathbf{L}$ (or matrix $\mathbf{H}$ if it contains more than one row) and conducts the hypothesis test contained in Equation 9.8. The ESTIMATE statement also informs the creation of the contrast coefficient vector $\mathbf{L}$ (or matrix $\mathbf{H}$ if it contains more than one row) and provides the estimate of the linear combination being tested in Equation 9.8.

```
PROC MIXED DATA=dset1 NOPROFILE;
   CLASS i j k;
   MODEL y=i/SOLUTION;
   RANDOM j i*j;
   CONTRAST 't1' i 1 -1;
   ESTIMATE 't1' i 1 -1;
RUN;
```

Output – Example 9.2:

Model Information	
Data Set	WORK.DSET1
Dependent Variable	y
Covariance Structure	Variance Components
Estimation Method	REML
Residual Variance Method	Parameter
Fixed Effects SE Method	Model-Based
Degrees of Freedom Method	Containment

Class Level Information		
Class	Levels	Values
i	4	1 2 3 4
j	5	1 2 3 4 5
k	3	1 2 3

Dimensions	
Covariance Parameters	3
Columns in X	5
Columns in Z	24
Subjects	1
Max Obs Per Subject	57

Number of Observations	
Number of Observations Read	57
Number of Observations Used	57
Number of Observations Not Used	0

Iteration History			
Iteration	Evaluations	-2 Res Log Like	Criterion
0	1	320.05312519	
1	3	220.11948188	0.00226234
2	1	219.96055340	0.00015973
3	1	219.95027211	0.00000104
4	1	219.95020819	0.00000000

Convergence criteria met.

Covariance Parameter Estimates	
Cov Parm	Estimate
j	17.5759
i*j	3.5702
Residual	1.3087

Fit Statistics	
-2 Res Log Likelihood	220.0
AIC (smaller is better)	226.0
AICC (smaller is better)	226.4
BIC (smaller is better)	224.8

Solution for Fixed Effects						
Effect	i	Estimate	Standard Error	DF	t Value	Pr > |t|
Intercept		39.8214	2.0776	4	19.17	<.0001
i	1	-29.5033	1.3656	11	-21.60	<.0001
i	2	-18.3269	1.2659	11	-14.48	<.0001
i	3	-9.1111	1.2659	11	-7.20	<.0001
i	4	0	.	.	.	.

Type 3 Tests of Fixed Effects				
Effect	Num DF	Den DF	F Value	Pr > F
i	3	11	173.67	<.0001

Estimates					
Label	Estimate	Standard Error	DF	t Value	Pr > |t|
t1	-11.1764	1.3656	11	-8.18	<.0001

Contrasts				
Label	Num DF	Den DF	F Value	Pr > F
t1	1	11	66.98	<.0001

The above sections of output result from the PROC MIXED step. The section of output entitled "Iteration History" provides the value of the objective function, *-2logREML*, for the different iterations, matching those of the IML optimization routine in the previous example.

Though the resulting values are the same for the two examples, the IML code is designed to work for the model and assumptions provided in these examples. The

MIXED procedure has many additional features that enhance its ability to analyze the linear mixed model. The IML code in the previous example is great for academic purposes and can be enhanced to include additional features if desired.

Though the section of output entitled "Contrasts" tests the same hypothesis as the hypothesis tested in the "Estimates" section of output, it need not. It can be shown that the F test statistic under "Contrasts" is the same as the square of the T test statistic under "Estimates". However, the ESTIMATE statement in PROC MIXED allows for the contrast coefficients to occupy only a one row vector (the row vector L in Equations 9.8 and 9.9). Whereas, the CONTRAST statement in PROC MIXED allows for the contrast coefficients to occupy multiple rows in a matrix (matrix **H** in Equations 9.12 and 9.13).

9.3 Chapter Exercises

9.1 Consider Examples 9.1 and 9.2. Enhance the code in Example 9.1 to include all the information contained in the section of output for Example 9.2 entitled "Solution for Fixed Effects"; including standard error, degrees of freedom, test statistic, and P value for each of the fixed effect parameter estimates. Use PROC MIXED to verify your answers.

9.2 Consider Examples 9.1 and 9.2. Enhance the code in Example 9.1 to include all the information contained in the section of output for Example 9.2 entitled "Fit Statistics". The formulas for the fit statistics can be found in the SAS Online Documentation under the heading "The MIXED Procedure: PROC MIXED Statement". Use PROC MIXED to verify your answers.

9.3 The following SAS code generates a set of mixed model data for analysis using PROC IML. The mixed model for this example can be defined as follows:

$$y_{ijk} = \mu + \alpha_i + b_j + \alpha b_{ij} + e_{ijk}; i = 1 \dots 3, j = 1 \dots 4, k = 1,2;$$

where

$$b_j \sim Normal(0, \sigma_b^2), \quad \sigma_b^2 = 4$$

$$\alpha b_{ij} \sim Normal(0, \sigma_{\alpha b}^2), \quad \sigma_{\alpha b}^2 = 9$$

$$e_{ijk} \sim Normal(0, \sigma_e^2), \quad \sigma_e^2 = 1$$

```
/** Mixed model data **/
DATA dset1;
    a = 3;
    b = 4;
    c = 2;
    ARRAY mu{3} (5 7 14); /** fixed factor A means **/
```

```
    sd_b=2; /** Standard deviation of factor B **/
    sd_ab=3; /** Standard deviation of A*B interaction **/
    sd_e=1; /**Standard deviation of random error **/
    DO j = 1 TO b;
        eb = RANNOR (0)*sd_b;
        DO i = 1 TO a;
            eab = RANNOR (0)*sd_ab;
            DO k = 1 TO c;
                y = mu{i} + eb + eab + RANNOR (0)*sd_e;
                /** response vector **/
                  mus=mu{i};
                OUTPUT;
            END;
        END;
    END;
RUN;

PROC SORT DATA=dset1;
    BY i j k;
RUN;
```

Analyze the data generated by the above code using IML. Include in the analysis a complete ANOVA table (including source, df, SS, MS, F, P value) for the fixed effects. For the random effects, include estimates of the covariance parameters. Repeat the analysis using PROC MIXED to verify the PROC IML computations.

9.4 The following SAS code generates a set of mixed model data for analysis using PROC IML. The mixed model for this example can be defined as follows:

$$y_{ijk} = \mu + \alpha_i + b_{j(i)} + e_{ijk}; i = 1,2; j = 1\ldots3; k = 1\ldots4;$$

where

$$b_{j(i)} \sim Normal(0, \sigma_b^2), \quad \sigma_b^2 = 4$$

$$e_{ijk} \sim Normal(0, \sigma_e^2), \quad \sigma_e^2 = 1$$

and $b_{j(i)}$ indicates the random factor B nested within the fixed factor A.

```
/** Mixed model data **/
DATA dset1;
    a = 2;
    b = 3;
    c = 4;
    ARRAY mu{3} (4 8 9); /** fixed factor A means **/
    sd_ab=2; /** Standard deviation of factor B(A) **/
```

```
sd_e=1; /** Standard deviation of random error **/
DO i = 1 TO a;
    DO j = 1 TO b;
        eab = RANNOR (0)*sd_ab;
        DO k = 1 TO c;
            y = mu{i} + eab + RANNOR (0)*sd_e; /**
            response vector **/
    OUTPUT;
            END;
        END;
    END;
RUN;
```

Analyze the data generated by the above code using IML. Include in the analysis a complete ANOVA table (including source, df, SS, MS, F, *P* value) for the fixed effects. For the random effects, include estimates of the covariance parameters. Repeat the analysis using PROC MIXED to verify the PROC IML computations.

9.5 The following SAS code generates a set of mixed model data for analysis using PROC IML. The mixed model for this example can be defined as follows:

$$y_{ijk} = \mu + a_i + b_{j(i)} + e_{ijk}; i = 1 \ldots 3; j = 1,2; k = 1,2;$$

where

$$a_i \sim Normal(0, \sigma_a^2), \quad \sigma_a^2 = 9$$

$$b_{j(i)} \sim Normal(0, \sigma_b^2), \quad \sigma_b^2 = 4$$

$$e_{ijk} \sim Normal(0, \sigma_e^2), \quad \sigma_e^2 = 1$$

This is called a "random coefficients" model where all the factors in the model are random.

```
/** Mixed model data **/
DATA dset1;
    a = 3;
    b = 4;
    c = 2;
    mu=5; /** overall mean or intercept **/
    sd_a=2; /** Standard deviation of factor A **/
    sd_ab=3; /** Standard deviation of A*B interaction **/
    sd_e=1; /**Standard deviation of random error **/
    DO i = 1 TO a;
        ea = RANNOR (0)*sd_a;/**
        DO j = 1 TO b;
```

```
        eab = RANNOR (0)*sd_ab;
        DO k = 1 TO c;
            y = mu + ea + eab + RANNOR (0)*sd_e;
            /** response vector **/
            OUTPUT;
        END;
    END;
  END;
RUN;
```

Analyze the data generated by the above code using IML. Include in the analysis estimates of the covariance parameters. Using Equation 9.13, conduct a hypothesis test testing

$$H_0 : \sigma_a^2 = 0 \text{ and } \sigma_b^2 = 0 \text{ vs.} H_A: \text{At least one of the variances} > 0$$

Using Equation 9.14, construct simultaneous 95% prediction intervals about σ_a^2 and σ_b^2.
Repeat the analysis using PROC MIXED to verify the PROC IML computations.

9.6 Refer to Exercise 9.3. Using IML test the hypothesis

$$H_0 : \alpha_1 = \alpha_2 \text{ vs.} H_A : \alpha_1 + \alpha_2$$

Repeat the hypothesis test in MIXED to verify your IML results.

Chapter 10
Statistical Computation Methods

Within the context of linear models, there are different algorithms that have been developed to make computations more accurate and/or efficient (fewer basic operations). One problem deals with the size of a matrix. When trying to construct an identity matrix with 50,000 rows and columns, using the I function, SAS returns an error message indicating it is "Unable to allocate sufficient memory" if there is not at least 2 Gigabytes of memory available for the operations. Similar results can occur while trying to construct, invert, and otherwise intensely manipulate other large matrices. The direct inversion of such matrices as the $X'X$ matrix can lead to rounding errors that reduce the numerical accuracy of all subsequent computations.

When memory, speed, and/or numerical accuracy are issues, researchers can increase their computational abilities or use different computational methods. SAS developers have gone to great lengths to increase the speed and accuracy of their software, and have done a very good job. With double-precision floating-point arithmetic, SAS offers a level of precision of about 16 decimal digits. As can be seen in the examples given in this chapter, there are no differences in the numeric results, regardless of the method employed (see footnote on page 108).

Regarding speed, there are differences due to the number of basic operations being performed by each method. However, most researchers will not complain if a method takes four seconds to complete rather than two seconds. The end of this chapter provides the results of a basic example in which each method was tried and the time recorded. The differences among most methods were minimal. Where the researcher may be more concerned is when conducting simulations in which a method is employed thousands of times. In that case, a method that takes two seconds rather than four seconds can significantly reduce the time needed for the simulations to complete.

This chapter demonstrates the analysis of a linear model using several different computational methods. The first example uses PROC GLM to conduct a basic ANOVA. The second example in this chapter uses the analytic formulas to demonstrate the same type of analysis. After the analysis using the analytic formulas, the remaining examples demonstrate different algorithms that were developed to eliminate the need for certain matrix inversions and to reduce the number of simple operations. The output of each program can be compared to show the reader that

J.J. Perrett, *A SAS/IML Companion for Linear Models*, Statistics and Computing, DOI 10.1007/978-1-4419-5557-9_10, © Springer Science+Business Media, LLC 2010

for this specific linear model and set of data the numeric results are numerically identical for all the different algorithms employed.

Consider the following linear means model:

$$y = \mu_1 + \mu_2 + \mu_3 + \varepsilon.$$

This model is a specific use of the more general form of the linear model defined in Equation 8.1.

Example 10.1 – The analysis of a linear model using GLM:

This first example uses the GLM procedure to analyze data.

```
DATA data1;
INPUT y x1 x2 x3;
DATALINES;
12   -1 -1 -1
101  -1 -1  1
23    1 -1  1
39   -1  1 -1
52    1  1 -1
43    1  1  1
91    1 -1 -1
4    -1  1  1
;
RUN;
TITLE "Method: GLM";
PROC GLM DATA=Data1;
    MODEL y=x1 x2 x3/SOLUTION NOINT;
    ESTIMATE 'kpB' x1 1 x2 0 x3 -3;
    CONTRAST 'x2=x3' x2 1,x3 -1;
QUIT;
```

Output – Example 10.1:

Method: GLM

The GLM Procedure

Dependent Variable: y

Source	DF	Sum of Squares	Mean Square	F Value	Pr > F
Model	3	1407.37500	469.12500	0.10	0.9575
Error	5	23837.62500	4767.52500		
Uncorrected Total	8	25245.00000			

The above section of output is the abbreviated ANOVA table. The MSE, 4767.525, can be found in that section.

Output – Example 10.1 (continued):

```
                              Method: GLM
                          The GLM Procedure
Contrast      DF    Contrast SS    Mean Square    F Value    Pr > F
x2=x3         2     1056.250000     528.125000       0.11    0.8973

                              Standard
Parameter        Estimate          Error     t Value    Pr > |t|
kpB            15.2500000       77.1971907       0.20     0.8512
                              Standard
Parameter        Estimate          Error     t Value    Pr > |t|
x1             6.62500000      24.41189515       0.27     0.7969
x2           -11.12500000      24.41189515      -0.46     0.6677
x3            -2.87500000      24.41189515      -0.12     0.9108
```

The above section of output includes the results of a linear contrast, a linear function of the parameter estimates, and parameter estimates. The contrast is a result of the CONTRAST statement and tests the hypothesis,

$$H_0: \mu_2 = \mu_3.$$

The ESTIMATE statement produces the estimate labeled "kpB" which computes and tests the estimate of the linear function

$$k'\beta = 1\mu_1 + 0\mu_2 - 3\mu_3.$$

Example 10.2 – The analysis of a linear model using the analytic formulas in IML:

This second example uses the IML procedure to analyze the linear model data from the previous example using the analytic formulas. The analytic formulas of interest in the analysis include the following:

$$\hat{\beta} = (X'X)^{-1}X'Y = X^-Y$$

$$\hat{\sigma}^2 = \frac{1}{n - Rank(X)}(Y'Y - Y'X\hat{\beta})$$

$$k'\hat{\beta}$$

$$SE(k'\hat{\beta}) = \hat{\sigma} \sqrt{k'(X'X)^{-1}k}$$

$$t = \frac{k'\hat{\beta} - k'\beta}{SE(k'\hat{\beta})}$$

$$F = \frac{(H\hat{\beta} - h)' \left[H(X'X)^{-1}H'\right]^{-1} (H\hat{\beta} - h)}{\hat{\sigma}^2(Rank(H))}$$

The following code directly implements the above analytic formulas.

```
TITLE "Method: Analytic Formulas";
PROC IML;
   x={-1 -1 -1,-1 -1 1,1 -1 1,-1 1 -1,1 1 -1,1 1 1,1 -1 -1,
      -1 1 1};
   y={12,101,23,39,52,43,91,4};
   n=NROW(x);
   p=NCOL(x);
   xpx=x'*x;
   xpxi=INV(xpx);
   xpy=x'*y;
   k={1,0,-3};
   START matrank(x); /** matrank computes matrix rank **/
      matrank=ROUND(TRACE(x*GINV(x)));
      RETURN (matrank);
   FINISH;
   /** Create Module: est to check for estimability **/
   START est(x,k);
      xk=x'||k;
      IF matrank(xk)^= matrank(x) | matrank(k)^= NCOL(k)
      THEN PRINT "Warning: Function" k "is not estimable!";
   FINISH;
   /** check for estimability of k **/
   CALL est(x,k);
   h={0 1 0,0 0 1};

   /** checking for estimability of h **/
   CALL est(x,h');
   lilh={0,0};
   b=GINV(x)*y; /** could also have used SOLVE(xpx,xpy); **/
   s2=(1/(n-p))*(y'*y-xpy'*b);
   cov_b=s2*xpxi;
   kpb=k'*b;
   se_kpb=SQRT(s2*(k'*xpxi*k));
```

```
    t_kpb=kpb/se_kpb;
    abs_t=ABS(t_kpb);
    p_val1=2*(1-PROBT(abs_t,n-p));
    ll=kpb-se_kpb*TINV(.975,n-p);
    ul=kpb+se_kpb*TINV(.975,n-p);
    q=matrank(h);
    f=((h*b-lilh)'*INV(h*xpxi*h')*(h*b-lilh))/(q*s2);
    p_val2=1-PROBF(f,q,n-p);
    PRINT n p b s2,kpb se_kpb t_kpb p_val1 ll ul,h lilh q f
        p_val2;
QUIT;
```

Output – Example 10.2:

```
            Method: Analytic Formulas
              n         p         b          s2
              8         3      6.625     4767.525
                               -11.125
                                -2.875
     kpb      se_kpb      t_kpb     p_val1           ll          ul
    15.25   77.197191   0.197546  0.8511801   -183.1917   213.6917


    h                         lilh         q         f      p_val2
    0           1        0        0         2  0.1107755  0.8972763
```

The resulting numbers in the above (analytic formulas) output are identical to those for the previous (GLM) output.

10.1 The Square Root Method

The Square Root method uses Cholesky's decomposition algorithm to compute matrices, vectors, and scalars useful in the analysis of linear models. This method is referred to as the "Square Root" method in literature (Graybill, 2000). The appeal of this method is that it avoids computing the inverse of the $X'X$ matrix. This computation can be time consuming, and, if ill-conditioned, can make it prohibitive to compute reasonable point estimates. This method is restricted to applications where $X'X$ is positive definite.

A basic application of the Square Root method uses the ROOT function defined in IML. The ROOT function uses Cholesky's decomposition algorithm to decompose a symmetric positive definite matrix. If the matrix Z is defined as

$$Z = [X|Y]$$

where X and Y come from Equation 8.1, then the matrix $Z'Z$ will have the following composition:

$$Z'Z = \begin{bmatrix} X'X & X'Y \\ Y'X & Y'Y \end{bmatrix}.$$

This matrix is symmetric. When it is positive definite, applying the ROOT function to it results in the matrix R with the following composition:

$$R = \text{ROOT}(Z'Z) = \begin{bmatrix} T & t \\ O & v \end{bmatrix}$$

The TRISOLV function in IML can be used to solve the equation $T\hat{\beta} = t$ for $\hat{\beta}$. Then s^2 can be computed using the equation

$$s^2 = \left(\frac{1}{n-p}\right)v^2.$$

Notice that for computing these two estimates, no matrices had to be inverted. This may save time and reduce rounding errors associated with inverting certain matrices.

Example 10.3 – The analysis of a linear model using the Square Root method in IML:

The following code uses the ROOT function in IML to conduct a data analysis using the Square Root method.

```
TITLE "Method: Square Root 1";
PROC IML;
    x={-1 -1 -1,-1 -1 1,1 -1 1,-1 1 -1,1 1 -1,1 1 1,1 -1 -1,
        -1 1 1};
    y={12,101,23,39,52,43,91,4};
    n=NROW (x);
    p=NCOL (x);
    xpx=x'*x;
    xpy=x'*y;
    z=x||y;
    zpz=z'*z;
    r=ROOT (zpz);
    t1=r [1:p,1:p];
    t2=r [1:p,p+1];
    v=r [p+1,p+1];
    b=TRISOLV (1,t1,t2);
    k={1,0,-3};
    START matrank(x); /** matrank computes matrix rank **/
        matrank=ROUND(TRACE(x*GINV(x)));
        RETURN (matrank);
```

```
      FINISH;
      /** Create Module: est to check for estimability **/
      START est(x,k);
        xk=x'||k;
        IF matrank(xk)^= matrank(x) | matrank(k)^= NCOL(k)
        THEN PRINT "Warning: Function" k "is not estimable!";
      FINISH;
      /** check for estimability of k **/
      CALL est(x,k);
      h={0 1 0,0 0 1};

   /** checking for estimability of h **/
   CALL est (x,h');
   lilh={0,0};
   s2=(1/(n-p))*v##2;
   cov_b=s2*t2'*t2;
   kpb=k'*b;
   invt1=INV (t1);
   xpxi=invt1*invt1';
   se_kpb=SQRT (s2*k'*xpxi*k);
   t_kpb=kpb/se_kpb;
   abs_t=ABS (t_kpb);
   p_val1=2*(1-PROBT(abs_t,n-p));
   ll=kpb-se_kpb*TINV (.975,n-p);
   ul=kpb+se_kpb*TINV (.975,n-p);
   q=matrank(h);
   f=((h*b-lilh)'*INV(h*xpxi*h')*(h*b-lilh))/(q*s2);
   p_val2=1-PROBF(f,q,n-p);
   PRINT n p b s2,kpb se_kpb t_kpb p_val1 ll ul, f p_val2;
QUIT;
```

The matrix $X'X$ is not directly inverted in the above code, but is computed by inverting the upper-triangular matrix **T1**.

Output – Example 10.3:

```
                    Method: Square Root 1
              n          p          b          s2
              8          3      6.625   4767.525
                              -11.125
                               -2.875
      kpb      se_kpb      t_kpb      p_val1           ll          ul
    15.25 77.197191   0.197546  0.8511801  -183.1917   213.6917
                               f    p_val2
                       0.1107755  0.8972763
```

A more complex use of the Square Root method is to augment matrices and vectors (side-by-side) used in traditional linear models analysis, and convert them into new matrices that can be used for simplified equations associated with the analysis of linear models (see Graybill 2000 and Kopitzke 1975). For example, construct the matrix A consisting of the following augmented matrices and vectors:

$$A = [X'X|X'Y|k_1|k_2|H']$$

Using this convention, $X'X$ and $X'Y$ come from the traditional normal equations,

$$X'X\hat{\beta} = X'Y.$$

The vectors k_1 and k_2 represent coefficient column vectors used to estimate the linear functions $k_1'\beta$ and $k_2'\beta$. The matrix H is associated with testing the hypothesis $H_0:H\beta = h$ versus the alternative $H_A:H\beta \neq h$. Performing Cholesky's decomposition algorithm on the matrix A results in the following matrix, R:

$$R = [T|t|a_1|a_2|G].$$

The dimensions of the matrices and vectors that make up the matrix, R, are of the same dimension as their counterpart in the matrix A. From the matrix R, the following equations can be used for an appropriate linear models analysis:

Formal Equations	Square Root Equations
$\hat{\sigma}^2 = \dfrac{1}{n - Rank(X)}(Y'Y - Y'X\hat{\beta})$	$\hat{\sigma}^2 = \dfrac{1}{n - Rank(X)}(Y'Y - t't)$
$k'\hat{\beta}$	$a_i't$
$SE(k'\hat{\beta}) = \hat{\sigma}\sqrt{k'(X'X)^{-1}k}$	$SE(k_i'\hat{\beta}) = \hat{\sigma}\sqrt{a_i'a_i}$
$COV(k_i'\hat{\beta}, k_j'\hat{\beta}) = \hat{\sigma}\left[k_i'(X'X)^{-1}k_j\right]$	$COV(k_i'\hat{\beta}, k_j'\hat{\beta}) = \hat{\sigma}(a_i'a_j)$
$CORR(k_i'\hat{\beta}, k_j'\hat{\beta}) = \dfrac{k_i'(X'X)^{-1}k_j}{\sqrt{k_i'(X'X)^{-1}k_i}\sqrt{k_j'(X'X)^{-1}k_j}}$	$CORR(k_i'\hat{\beta}, k_j'\hat{\beta}) = \dfrac{a_i'a_j}{\sqrt{a_i'a_i}\sqrt{a_j'a_j}}$
$t = \dfrac{k'\hat{\beta} - k'\beta}{SE(k'\hat{\beta})}$	$t = \dfrac{a_i't - a_0}{SE(k_i'\hat{\beta})}$
$F = \dfrac{(H\hat{\beta} - h)'\left[H(X'X)^{-1}H'\right]^{-1}(H\hat{\beta} - h)}{\hat{\sigma}^2(Rank(H))}$	$F = \dfrac{t_0't_0}{\hat{\sigma}^2(Rank(H))}$

t_0 is found by applying the Square Root method to the following matrix, B:

$$\text{Original Matrix: } B = [GG'|g]$$

After Square Root method is applied: $C = [T_0|t_0]$. The following three examples demonstrates an application of this second square root method using the IML procedure. The purpose of this example is to show how the code could be written. The ESTIMATE statement produces the estimate labeled "kpB" which computes and tests the estimate of the linear function

$$k'\beta = 1\mu_1 + 0\mu_2 - 3\mu_3.$$

Example 10.4 – The analysis of a linear model using the second Square Root method in IML:

This example uses the IML procedure to analyze the linear model data using the second Square Root method. In this example the Cholesky decomposition algorithm equations are coded into the IML module SQROOT. This is necessary because the IML function ROOT requires a square input matrix and the augmented matrix A is not a square matrix. The module SQROOT is able to perform the Cholesky decomposition algorithm on the augmented matrix A.

```
TITLE "Method: Square Root 2";
PROC IML;
    x={-1 -1 -1,-1 -1 1,1 -1 1,-1 1 -1,1 1 -1,1 1 1,1 -1 -1,
        -1 1 1};
    y={12,101,23,39,52,43,91,4};
    n=NROW(x);
    p=NCOL(x);
    xpx=x'*x;
    xpy=x'*y;
    k={1,0,-3};
    START matrank(x); /** matrank computes matrix rank **/
        matrank=ROUND(TRACE(x*GINV(x)));
        RETURN (matrank);
    FINISH;
    /** Create Module: est to check for estimability **/
    START est(x,k);
        xk=x'||k;
        IF matrank(xk)^= matrank(x) | matrank(k)^= NCOL(k)
        THEN PRINT "Warning: Function" k "is not estimable!";
    FINISH;
    /** check for estimability of k **/
    CALL est(x,k);
    h={0 1 0,0 0 1};

    /** checking for estimability of h **/
    CALL est(x,h');
    lilh={0,0};
    a=xpx||xpy||k||h';
    START sqroot(a); /** Cholesky's Decomposition on an mxn
    matrix **/
```

```
    rows=NROW(a);
    cols=NCOL(a);
    r=J(rows,cols,0);
    r[1,1]=SQRT(a[1,1]);
    r[1,]=a[1,]/r[1,1];
    DO i=2 TO rows;
        r[i,i]=SQRT(a[i,i]-SSQ(r[1:i-1,i]));
        r[i,i:cols]= (a[i,i:cols]-r[1:i-1,i]'
                      *r[1:i-1,i:cols])/r[i,i];
    END;
    RETURN(r);
FINISH sqroot;
r=sqroot(a);
t=r[1:p,1:p];     /** Define the matrix T **/
lilt=r[1:p,p+1];   /** Define the vector t **/
a=r[1:p,p+2];     /** Define the vector a **/
g=(r[1:p,p+3:p+4])'; /** Define the matrix G **/
lilg=g*lilt-lilh;  /**Define the vector g **/
w=g*g'||lilg;
c=sqroot(w);
t0=c[1:NROW(lilh),1:NROW(lilh)];
lilt0=c[1:NROW(lilh),NROW(lilh)+1];
b=SOLVE(t,lilt);
s2=(1/(n-p))*(y'*y-lilt'*lilt);
cov_b=s2*lilt'*lilt;
kpb=a'*lilt;
se_kpb=SQRT(s2*a'*a);
t_kpb=kpb/se_kpb;
abs_t=ABS(t_kpb);
p_val1=2*(1-PROBT(abs_t,n-p));
ll=kpb-se_kpb*TINV(.975,n-p);
ul=kpb+se_kpb*TINV(.975,n-p);
q=matrank(h);
f=(lilt0'*lilt0)/(q*s2);
p_val2=1-PROBF(f,q,n-p);
PRINT n p b s2,kpb se_kpb t_kpb p_val1 ll ul,h lilh
    q f p_val2;
QUIT;
```

Output – Example 10.4:

Method: Square Root 2

n	p	b	s2
8	3	6.625	4767.525
		-11.125	
		-2.875	

kpb	se_kpb	t_kpb	p_val1	ll	ul
15.25	77.197191	0.197546	0.8511801	-183.1917	213.6917

h			lilh	q	f	p_val2
0	1	0	0	2	0.1107755	0.8972763
0	0	1	0			

The resulting numbers in the above (analytic formulas) output are identical to those for the GLM output. The Square Root methods are intended to reduce rounding error in certain cases and to increase speed. The rounding errors result from the calculation of the inverses. Because the second Square Root method does not require the use of inverses of matrices, the associated rounding errors are avoided.

10.2 The Sweep Method

The Sweep method, which makes use of the Sweep algorithm (see Goodnight 1978 and Hocking 2003), is a method of calculating the matrices

$$(X'X)^{-1},$$

$$(X'X)^{-1}X'Y,$$

and

$$Y'Y - Y'X(X'X)^{-1}X'Y.$$

The Sweep algorithm is faster than the analytic method in computing sums of squares and the least squares estimates to the linear models parameters because it computes these values without having to invert matrices (Goodnight, 1979). Define the matrix $Z = [X|Y]$ where X and Y come from Equation 8.1, then the matrix $Z'Z$ will have the following composition:

$$Z'Z = \begin{bmatrix} X'X & X'Y \\ Y'X & Y'Y \end{bmatrix}$$

If $X'X$ is a $q \times q$ positive definite matrix then sweeping the matrix $Z'Z$ by the first q pivots (diagonal elements) will result in the following matrix, S:

$$S = \text{SWEEP}(Z'Z, \{1,2,\ldots,q\}) = \begin{bmatrix} (X'X)^{-1} & (X'X)^{-1}X'Y \\ -[(X'X)^{-1}X'Y] & Y'Y - Y'X(X'X)^{-1}X'Y \end{bmatrix}$$

From the "swept" matrix, S, the following can easily be computed:

$$\hat{\beta} = (X'X)^{-1}X'Y$$

$$\hat{\sigma}^2 = \frac{1}{n - Rank(X)} \left[Y'Y - Y'X(X'X)^{-1}X'Y \right]$$

$$k'\hat{\beta}$$

$$SE(k'\hat{\beta}) = \hat{\sigma}\sqrt{k'(X'X)^{-1}k}$$

$$COV(k_i'\hat{\beta}, k_j'\hat{\beta}) = \hat{\sigma}\left[k_i'(X'X)^{-1}k_j \right]$$

$$CORR(k_i'\hat{\beta}, k_j'\hat{\beta}) = \frac{k_i'(X'X)^{-1}k_j}{\sqrt{k_i'(X'X)^{-1}k_i}\sqrt{k_j'(X'X)^{-1}k_j}}$$

$$t = \frac{k'\hat{\beta} - k'\beta}{SE(k'\hat{\beta})}$$

$$R(\beta_1|\beta_2)$$

$$F = \frac{(H\hat{\beta} - h)\left[H(X'X)^{-1}H'\right]^{-1}(H\hat{\beta} - h)}{\hat{\sigma}^2(Rank(H))}$$

etc.

One benefit of the Sweep algorithm is that sweeping is reversible and can be done in segments that are also reversible. For example, if B=SWEEP(A,1) then the matrix B is the matrix A swept by the first pivot. Because of the reversibility property, A=SWEEP(B,1), meaning that sweeping a second time by the same pivot "unsweeps" the matrix. Also,

```
SWEEP (A, {1 2})  = SWEEP (SWEEP (A,{1}),{2})
                  = SWEEP (SWEEP (A, {1 2 3}), {3}).
```

This facility can be used to reduce redundancy, which also helps makes the Sweep method fast. The next example demonstrates an application of the Sweep algorithm using the IML procedure.

Example 10.5 – The Sweep method:

This example uses the Sweep method to perform the analysis of previous three examples. This example is designed to show how the code can be written.

```
PROC IML;
    x={-1 -1 -1,-1 -1 1,1 -1 1,-1 1 -1,1 1 -1,1 1 1,1 -1 -1,
       -1 1 1};
    y={12,101,23,39,52,43,91,4};
    n=NROW(x);
    p=NCOL(x);
```

```
k={1,0,-3};
START matrank(x); /** matrank computes matrix rank **/
   matrank=ROUND(TRACE(x*GINV(x)));
   RETURN (matrank);
FINISH;
/** Create Module: est to check for estimability **/
START est(x,k);
   xk=x`||k;
   IF matrank(xk)^= matrank(x) | matrank(k)^= NCOL(k)
   THEN PRINT "Warning: Function" k "is not estimable!";
FINISH;
/** check for estimability of k **/
CALL est(x,k);
h={0 1 0,0 0 1};

/** checking for estimability of h **/
CALL est(x,h`);
lilh={0,0};
xy=x||y;
z=xy`*xy;
sw=SWEEP(z,{1 2 3}); /** sw is z swept by all three pivots **/
xpxi=sw[1:p,1:p];
b=sw[1:p,p+1];
sse=sw[p+1,P+1]; /** SSE=R(1,2,3) **/
s2=(1/(n-p))*sse;
cov_b=s2*xpxi;
kpb=k`*b;
se_kpb=SQRT(s2*(k`*xpxi*k));
t_kpb=kpb/se_kpb;
abs_t=ABS(t_kpb);
p_val1=2*(1-PROBT(abs_t,n-p));
ll=kpb-se_kpb*TINV(.975,n-p);
ul=kpb+se_kpb*TINV(.975,n-p);
q=1;
sw1a=SWEEP(z,{2 3});
sw1b=SWEEP(sw,1);
ss1=sw1a[p+1,p+1]; /** SS1=R(1) **/
ssx1=ss1-sse; /** SSx1=R(1|2,3) **/
f1=(ssx1/sse)*((n-p)/(1));
sw2=SWEEP(sw,2);
ss2=sw2[p+1,p+1]; /** SS2=R(2) **/
ssx2=ss2-sse; /** SSx2=R(2|1,3) **/
f2=(ssx2/sse)*((n-p)/(1));
sw3=SWEEP(sw,3);
ss3=sw3[p+1,p+1]; /** SS3=R(3) **/
```

```
ssx3=ss3-sse; /** SSx3=R(3|1,2) **/
f3=(ssx3/sse)*((n-p)/(1));
q=matrank(h);
f=((h*b-lilh)'*INV(h*xpxi*h')*(h*b-lilh))/(q*s2);
p_val2=1-PROBF(f,q,n-p);
PRINT n p b s2,kpb se_kpb t_kpb p_val1 ll ul,h lilh q f
    p_val2,swla swlb,ssx1 ssx2 ssx3 f1 f2 f3;
QUIT;
```

Output – Example 10.5:

n	p	b	s2
8	3	6.625	4767.525
		-11.125	
		-2.875	

kpb	se_kpb	t_kpb	p_val1	ll	ul
15.25	77.197191	0.197546	0.8511801	-183.1917	213.6917

h		lilh		q	f	p_val2
0	1	0	0	2	0.1107755	0.8972763
0	0	1	0			

The results are identical to those found using the Analytic formulas (Example 10.2), the Square Root methods (Examples 10.3 and 10.4), and the Sweep method. With large data matrices the analysis using the Sweep method will be faster than the analysis using the analytic formulas because the Sweep method requires fewer basic operations than the analytic formulas.

Output – Example 10.5 (continued):

ssx1	ssx2	ssx3	f1	f2	f3
351.125	990.125	66.125	0.0736493	0.2076811	0.0138699

The above output lists the Type 3 sums of squares f statistics for testing the effects of the model parameters $\beta_1=$**X1**, $\beta_2=$**X2**, $\beta_3=$**X3** computed using the Sweep method. They are similar to results observed in the GLM output (Example 10.1).

Output – Example 10.5 (continued):

	swla		
8	0	0	53
0	0.125	0	-11.125
0	0	0.125	-2.875
53	11.125	2.875	24188.75
	swlb		
8	0	0	53
0	0.125	0	-11.125

| 0 | 0 | 0.125 | -2.875 |
| 53 | 11.125 | 2.875 | 24188.75 |

The above output demonstrates the reversibility of the Sweep method.

```
sw1a=SWEEP(z,{2 3});
sw1b=SWEEP(sw,1);
```

Additionally, it can be shown that

```
sw1b = SWEEP(SWEEP(z,{1 2 3}),1);
```

Both **SW1A** and **SW1B** are identical matrices used in computing the Type 3 sums of squares for testing the effect of β_1 given β_2 and β_3, R(x1|x2,x3). However, **SW1B** takes advantage of the fact that **Z** has already been swept by all three pivots in order to compute **SSE**. So, **SW1B** sweeps by only one pivot where **SW1A** sweeps by two. The result is increased speed and efficiency.

10.3 The QR Factorization Method

As with the Sweep and Square root methods, the QR factorization method (QR method) aims to analyze a linear model without using the analytic formula to compute $(X'X)^{-1}$. As mentioned before, using the analytic formula to compute $(X'X)^{-1}$ can lead to reduced numerical accuracy (rounding errors) and inefficiency (a lot of time to invert that matrix). The QR method uses QR factorization to factor the $n \times p$ design matrix X as $X = QR$ where Q is an $n \times p$ orthogonal matrix such that $Q'Q = I_p$, and R is a $p \times p$ upper triangular matrix such that the elements below the diagonal are all zeros.

Consider the model in Equation 8.1. The LS estimate of β is

$$\hat{\beta} = (X'X)^{-1}X'Y.$$

Next, apply the results of the QR factorization to this equation:

$$\hat{\beta} = [(QR)'(QR)]^{-1}(QR)'Y$$
$$=[R'Q'QR]^{-1}R'Q'Y$$
$$=[R'R]^{-1}R'Q'Y$$
$$=R^{-1}(R')^{-1}R'Q'Y$$
$$=R^{-1}Q'Y$$

(Weisberg, 2005) SAS/IML contains a QR Call that will conduct the factorization. Additionally, the QR Call can produce the $Q'Y$ column vector. Also, the TRISOLV function in IML is designed to solve linear systems that involve triangular matrices. The combination of the QR Call and the TRISOLV function computes the LS estimate of β in an efficient and accurate manner.

Example 10.6 – The QR method:

This example uses the QR method to perform the analysis of previous four examples. This example is designed to show how the code can be written.

```
PROC IML;
    x={-1 -1 -1,-1 -1 1,1 -1 1,-1 1 -1,1 1 -1,1 1 1,1 -1 -1,
        -1 1 1};
    y={12,101,23,39,52,43,91,4};
    START matrank(x); /** matrank computes matrix rank **/
        matrank=ROUND(TRACE(x*GINV(x)));
        RETURN (matrank);
    FINISH;
    n=NROW(x);
    p=NCOL(x);
    k={1,0,-3};
    /** Create Module: est to check for estimability **/
    START est(x,k);
        xk=x'||k;
        IF matrank(xk)^= matrank(x) | matrank(k)^= NCOL(k)
        THEN PRINT "Warning: Function" k "is not estimable!";
    FINISH;
    /** check for estimability of k **/
    CALL est(x,k);
    CALL QR(qpy,r,piv,lindep,x, y); /** Compute QPY **/
    IF lindep=0 THEN b=TRISOLV(1,r,qpy[1:p],piv);
    CALL QR(q,r,piv,lindep,x); /** Compute Q for computing H **/
    q=q[,1:p];
    h=q*q';
    SSE=y'*(I(n)-h)*y;
    rinv=INV(r);
    xpxi=rinv*rinv';
    s2=(1/(n-p))*sse;
    cov_b=s2*xpxi;
    kpb=k'*b;
    se_kpb=SQRT(s2*(k'*xpxi*k));
    t_kpb=kpb/se_kpb;
    abs_t=ABS(t_kpb);
    p_val1=2*(1-PROBT(abs_t,n-p));
    ll=kpb-se_kpb*TINV(.975,n-p);
    ul=kpb+se_kpb*TINV(.975,n-p);
    PRINT n p b s2,kpb se_kpb t_kpb p_val1 ll ul;
QUIT;
```

The first Call QR statement is used to compute $Q'Y$. The second Call QR statement is used to compute the matrix Q. The $p \times n$ matrix $Q'Y$ can be much smaller than the $n \times n$ matrix Q. Consequently, when the quantity $Q'Y$ is needed, as when computing the estimate $\hat{\beta}$, it can be more efficient to compute $Q'Y$ directly using the Call QR statement rather than using the Call QR statement to compute Q and then using matrix algebra to get $Q'Y$. The $Q'Y$ matrix is then used in the TRISOLV function to solve for the estimate $\hat{\beta}$. The IF-THEN portion of the statement checks for the matrix **R** to be nonsingular—a requirement of the TRISOLV function. The matrix **Q**, computed from the second Call QR statement, is then used to compute the matrix **H**, known as the *hat* matrix. This matrix is used for a variety of linear models statistics including many related to multiple linear regression analysis. One important application of the matrix **H** is in the computation of **SSE** as seen in the above code. The **XPXI** matrix is a computation of $(X'X)^{-1}$ using the inverse of the matrix **R**. The remaining computations in the program are similar to those in the previous examples. The numbers in the output match those of the previous examples.

Output – Example 10.6:

```
            n            p            b            s2
            8            3        6.625     4767.525
                                -11.125
                                 -2.875
      kpb      se_kpb      t_kpb      p_val1          ll           ul
    15.25   77.197191   0.197546  0.8511801  -183.1917    213.6917
```

As can be seen, there are multiple ways of avoiding the direct analytic computation of $(X'X)^{-1}$. When analyzing large data sets, these methods can save time over the analytic method and be more numerically accurate.

There is less than one second difference in the time required for the computations for each method to compute the least squares estimate of the model's mean parameter vector. The GLM procedure constructs many statistics and conducts a variety of tests. Consequently, it can take more time than other methods presented in this chapter.

A basic example was created to identify time differences among the different methods. The VNORMAL function in IML (independent Normally-distributed values based on mean=0 and variance=1) was used to generate a data set with 501 variables (1 column for **Y** and 500 columns for **X**) and 10,000 observations. Each method was then applied to the data set (separate PROC steps) to compute the least squares estimate, $\hat{\beta}$. The CPU time from the SAS LOG was recorded for each PROC step. When increasing the dimension of the design matrix, **X**, to 10,000 rows and 500 columns the Analytic approach took about 5 seconds. Using PROC GLM took about 3 seconds. The Square Root methods took about 2 seconds and the Sweep method about 2 seconds. Interestingly, the QR method took about 35 seconds, demonstrating that the QR Call in SAS may remove the necessity of inverting the $X'X$ matrix, but does not appear to improve on speed—at least not in this general situation. Though speed may not be an issue with a single analysis, researchers may want to incorporate thousands of similar analyses in a simulation study. It would then be of interest to choose a method that best fits the researchers needs and is at least as fast as other possible methods. Of the methods presented in this chapter, the Analytic method and the QR method appear to be slower methods. There are several factors that can influence the speed of the different methods, including the data, the additional statistics to be computed and tests to be conducted, as well as the amount and nature of the output. This small "study" did not take that information into account.

The different methods presented in this chapter have different benefits as well as different assumptions. It is highly recommended that the careful student of linear models applications study not only the analytic formulas that are used so often for proofs and theorems in theory courses, but also the computational efficiencies described in this chapter.

10.4 Chapter Exercises

10.1 Analyze the data in Example 8.2 using the Square Root Method in IML. Include in the analysis a complete ANOVA table. Repeat the analysis in GLM and compare the results to the IML results for accuracy.

10.2 Analyze the data in Example 8.2 using the Sweep Method in IML. Include in the analysis a complete ANOVA table. Repeat the analysis in GLM and compare the results to the IML results for accuracy.

10.3 Analyze the data in Example 8.2 using the QR Factorization Method in IML. Include in the analysis a complete ANOVA table. Repeat the analysis in GLM and compare the results to the IML results for accuracy.

10.4 Analyze the data in Example 8.6 using the Square Root Method in IML. Include in the analysis a complete ANOVA table. Repeat the analysis in GLM and compare the results to the IML results for accuracy.

10.5 Analyze the data in Example 8.6 using the Sweep Method in IML. Include in the analysis a complete ANOVA table. Repeat the analysis in GLM and compare the results to the IML results for accuracy.

10.6 Analyze the data in Example 8.6 using the QR Factorization Method in IML. Include in the analysis a complete ANOVA table. Repeat the analysis in GLM and compare the results to the IML results for accuracy.

10.7 Analyze the data in Example 8.9 using the Square Root Method in IML. Include in the analysis a complete ANOVA table. Repeat the analysis in GLM and compare the results to the IML results for accuracy.

10.8 Analyze the data in Example 8.9 using the Sweep Method in IML. Include in the analysis a complete ANOVA table. Repeat the analysis in GLM and compare the results to the IML results for accuracy.

10.9 Analyze the data in Example 8.9 using the QR Factorization Method in IML. Include in the analysis a complete ANOVA table. Repeat the analysis in GLM and compare the results to the IML results for accuracy.

10.10 Compare the Square Root, Sweep, and QR Factorization methods. Describe the differences in how they go about improving efficiency in the analysis of a linear model.

In Summary

The IML procedure can be very helpful in quickly and easily conducting small scale Monte Carlo-type simulation studies. This feature can be beneficial in a variety of research and academic situations. Also, there are many functions set up in IML for various tools. Among those tools are financial equations, wavelet functions, etc. As the theory and methodology associated with the linear model continue to develop, future editions of this book will explore these features. Meanwhile, descriptions and examples can be found in the *SAS/IML User's Guide, Version 9* and on the online help.

The SAS/STAT package includes canned procedures such as GLM, MIXED, and REG. Each of these procedures is designed to expedite statistical data analyses with built-in efficiencies that allow for high precision of complex computations with faster speeds. SAS/IML does not replace SAS/STAT. As can be seen from the previous examples, less code is required to analyze data in the context of linear models using the SAS/STAT procedures. Also, the SAS/STAT procedures are more efficient and numerically accurate. Also, the SAS/STAT procedures make available a plethora of output related to the analysis. A similar argument holds true for other procedures in SAS as well. However, SAS/IML has many benefits. Because SAS/IML can require definition of all the matrices and computations of an analysis, the user must know all the necessary formulas. This makes IML ideal for theoretical, applied, and hybrid linear models courses.

Analysis with SAS/STAT procedures is partially limited to the computations included in the procedure. Analysis with SAS/IML can be used to confirm new theory and research new methodologies. This opens the door for research using computations that are not currently contained in a SAS/STAT procedure. This may involve both computing final equations in Proc IML after prior steps have been performed using another procedure. This may also involve doing an entire analysis in the IML environment. For example, consider linear models analyses using the MIXED procedure. When desired, a SATTERTH or KENWARDROGER option may be used to approximate the degrees of freedom. However, suppose the user is researching a different method for computing degrees of freedom. Sums of squares and other summary information can be computed in the MIXED procedure. That information might then be used in an IML procedure where matrix operations can then be used to estimate the alternate degrees of freedom and consequent test results.

J.J. Perrett, *A SAS/IML Companion for Linear Models*, Statistics and Computing, DOI 10.1007/978-1-4419-5557-9, © Springer Science+Business Media, LLC 2010

References

Goodnight, James H. (1978), *The Sweep Operator: Its Importance in Statistical Computing*, SAS Technical Report Series, R-106, Cary, NC: SAS Institute Inc.

Goodnight, James H. (1979), "A Tutorial on the SWEEP Operator". *The American Statistician*. Vol. 33. No. 3: 149–158

Graybill, Franklin A. (2001), *Matrices with Applications in Statistics, Second Edition*, Belmont, CA: Duxbury Press.

Graybill, Franklin A. (2000), *Theory and Application of the Linear Model*, Belmont, CA: Duxbury Press.

Henderson, C.R. (1984), Applications of Linear Models in Animal Breeding, University of Guelph.

Hocking, Ronald R. (2003), *Methods and Applications of Linear Models: Regression and Analysis of Variance*, New York, NY: John Wiley & Sons Inc.

Kopitzke, R., Boardman, T. J., and Graybill, F. A. (1975), *Least Squares Programs – A Look at the Square Root Procedure*, American Statistician, Vol. 29, No. 1: 64–66

Littell, Ramon C., Milliken, George A., Stroup, Walter W., and Wolfinger, Russell D. (2006), *SAS System for Mixed Models, Second Edition*, Cary, NC: SAS Institute Inc.

SAS Institute Inc. (2009), *SAS/IML 9.2 User's Guide*, Cary, NC: SAS Institute Inc.

SAS Institute Inc. (2004), *SAS/STAT 9.1 User's Guide*, Cary, NC: SAS Institute Inc.

Thisted, Ronald A. (1988), *Elements of Statistical Computing: Numerical Computation*, London, England: Chapman and Hall Ltd.

Van Vleck, L.D. (1998) "Charles Roy Henderson, 1911-1989: a brief biography". *Journal of Animal Science*. 76(12):2959–61

Weisberg, Sanford. (2005), *Applied Linear Regression, Third Edition*, Hoboken, NJ: John Wiley & Sons.

Wolfinger, R.D., Tobias, R.D., and Sall, J. (1991), *Mixed Models: A Future Direction*, Proceedings of the Sixteenth Annual SAS Users Group Conference, SAS Institute Inc., Cary, NC, 1380–1388.

Index

SAS for Data Analysis

Intermediate Statistical Methods

Mervyn G. Marasinghe
William J. Kennedy

Content: Introduction to SAS language.- More on SAS programming and some applications.- Statistical graphics using SAS/GRAPH.- Statistical analysis of regression models.- Analysis of variance models.- Analysis of variance: random mixed effects models.- Appendices.- References.- Index.

2008.XII, 557 p. With 100 SAS Programs. Hardcover
Statistics and Computing
ISBN: 978-0-387-77371-1

R for SAS and SPSS Users

Robert A. Muenchen

Content: Introduction. The five main parts of SAS and SPSS.- Programming conventions.- Typographic conventions.- Installing & updating R.- Running R.- Help and documentation.- Programming language basics.- Data Acquisition.- Selecting Variables - Var, Variables.- Selecting observations - where, if select if, filter.- Selecting both variables and observations.- Converting data structures.- Data management.- Recoding variables. Value labels or formats (& measurement level).- Variable labels.- Generating data.- How R stores data.- Managing your files and workspace.- Graphics overview.- Traditional graphics.- The ggplot2 package.- Statistics.- Summary.- Conclusion.- Appendix A.- Appendix B.- Appendix C.- Bibliography.

2009. XVII, 470 p. Hardcover
Statistics and Computing
ISBN: 978-0-387-09417-5

Linear and Generalized Linear Mixed Models and Their Applications

Jiming Jiang

Content: Linear mixed models: Part I.- Linear mixed models: Part II.- Generalized linear mixed models: Part I.- Generalized linear mixed models: Part II.

2007. XIV, 257 p. Hardcover
Springer Series in Statistics
ISBN: 978-0-387-47941-5
